Deep Learning

This book focuses on deep learning (DL), which is an important aspect of data science, that includes predictive modeling. DL applications are widely used in domains such as finance, transport, healthcare, automanufacturing, and advertising. The design of the DL models based on artificial neural networks is influenced by the structure and operation of the brain. This book presents a comprehensive resource for those who seek a solid grasp of the techniques in DL.

Key features:

- Provides knowledge on theory and design of state-of-the-art deep learning models for real-world applications.
- Explains the concepts and terminology in problem-solving with deep learning.
- Explores the theoretical basis for major algorithms and approaches in deep learning.
- Discusses the enhancement techniques of deep learning models.
- Identifies the performance evaluation techniques for deep learning models.

Accordingly, the book covers the entire process flow of deep learning by providing awareness of each of the widely used models. This book can be used as a beginners' guide where the user can understand the associated concepts and techniques. This book will be a useful resource for undergraduate and postgraduate students, engineers, and researchers, who are starting to learn the subject of deep learning.

Dulani Meedeniya is a Professor in Computer Science and Engineering at the University of Moratuwa, Sri Lanka. She holds a PhD in Computer Science from the University of St Andrews, United Kingdom. She is the director of the Bio-Health Informatics group at her department and engages in a number of collaborative research projects. She is a co-author of 100+ publications in indexed journals, peer-reviewed conferences, and book chapters. Prof. Dulani has received several awards and grants for her contribution to research. She serves as a reviewer, program committee, and editorial team member in many international conferences and journals. Her main research interests are deep learning, software modeling and design, bio-health informatics, and technology-enhanced learning. She is a Fellow of HEA (UK), MIET, Senior Member of IEEE, Member of ACM, and a Chartered Engineer registered at EC (UK).

Deep Learning
A Beginners' Guide

Dulani Meedeniya

CRC Press
Taylor & Francis Group
Boca Raton London New York

CRC Press is an imprint of the
Taylor & Francis Group, an **informa** business

Designed cover image: Pdusit, Shutterstock Illustrator

First edition published 2024
by CRC Press
6000 Broken Sound Parkway NW, Suite 300, Boca Raton, FL 33487-2742

and by CRC Press
4 Park Square, Milton Park, Abingdon, Oxon, OX14 4RN

CRC Press is an imprint of Taylor & Francis Group, LLC

© 2024 Dulani Meedeniya

ISBN: 9781032473246 (hbk)
ISBN: 9781032487960 (pbk)
ISBN: 9781003390824 (ebk)

DOI: 10.1201/9781003390824

Typeset in Times
by Newgen Publishing UK

Contents

Preface ix
Acknowledgements xi
List of Abbreviations xiii

Chapter 1 Introduction 1
    1.1   Data-Driven Decision-Making and Society 1
    1.2   Overview of Deep Learning 2
    1.3   Bias and Variance 6
        1.3.1   Skewness of Data 7
        1.3.2   Bias 7
        1.3.3   Variance 8
        1.3.4   Trade-off Between Bias and Variance 8
    1.4   Supervised and Unsupervised Learning 11
    1.5   Supportive Tools and Libraries 12
        1.5.1   TensorFlow 13
        1.5.2   Keras 13
        1.5.3   PyTorch 14
        1.5.4   Jupyter Notebook 14
        1.5.5   NumPy and Pandas 14
        1.5.6   Tensor Hub 14
    Review Questions 15

Chapter 2 Concepts and Terminology 16
    2.1   Understanding Neural Networks 16
    2.2   Regression 18
        2.2.1   Linear Regression 19
        2.2.2   Logistic Regression 20
        2.2.3   Other Regression Methods 20
    2.3   Classification 21
    2.4   Hyperparameters 22
        2.4.1   Overview 22
        2.4.2   Weight Initialization 22
        2.4.3   Activation Function 24
        2.4.4   Learning Rate 29
        2.4.5   Loss Function 29
        2.4.6   Other Hyperparameters 32
    2.5   Model Training 33
        2.5.1   Model Selection 33
        2.5.2   Model Convergence 33

2.5.3   Overfitting and Underfitting 34
2.5.4   Regularization 37
2.5.5   Network Gradients 38
Review Questions 41

Chapter 3   State-of-the-Art Deep Learning Models: Part I 42

3.1   Overview of Neural Networks 42
3.2   Artificial Neural Networks 43
3.3   Recurrent Neural Network (RNN) 45
3.4   Convolutional Neural Networks 48
        3.4.1   Overview of Convolutional Neural Network 48
        3.4.2   Concepts of CNN 49
        3.4.3   Convolutional Layer 52
        3.4.4   Pooling Layer 54
        3.4.5   Fully Connected Layer 55
3.5   Comparison of ANN, RNN, and CNN 56
Review Questions 58

Chapter 4   State-of-the-Art Deep Learning Models: Part II 59

4.1   Feed-Forward Neural Network 59
4.2   Multi-layer Perceptrons 61
4.3   Generative Adversarial Network (GAN) 62
4.4   Variations of CNNs 64
        4.4.1   Residual Networks (ResNet) 64
        4.4.2   Inception Model 66
        4.4.3   GoogLeNet 67
        4.4.4   Xception Model 69
        4.4.5   DenseNet Model 69
        4.4.6   MobileNet Model 70
        4.4.7   VGG Model 71
        4.4.8   Comparison of CNN Architectures 71
4.5   Capsule Network 73
4.6   Autoencoders 76
4.7   Transformers 78
Review Questions 83

Chapter 5   Advanced Learning Techniques 84

5.1   Transfer Learning 84
        5.1.1   Overview of Transfer Learning 84
        5.1.2   Transfer Learning Process 85
        5.1.3   Transfer Learning Types, Categories,
                     and Strategies 87
        5.1.4   Transfer Learning Applications 89
        5.1.5   Transfer Learning Challenges 90

5.2   Reinforcement Learning 91
        5.2.1   Overview of Reinforcement Learning........................91
        5.2.2   Reinforcement Learning Process................................91
        5.2.3   Implementation and Scheduling Types 94
        5.2.4   Applications of Reinforcement Learning..................95
        5.2.5   Challenges of Reinforcement Learning.....................95
5.3   Federated Learning 96
        5.3.1   Overview of Federated Learning...............................96
        5.3.2   Federated Learning Process.......................................97
        5.3.3   Types and Properties of Federated Learning...........100
        5.3.4   Applications of Federated Learning........................101
        5.3.5   Challenges of Federated Learning...........................102
5.4   Multi-modeling with Ensemble Learning...............................103
        5.4.1   Overview of Ensemble Learning.............................103
        5.4.2   Ensemble Learning Process.....................................103
        5.4.3   Ensemble Learning Techniques...............................106
        5.4.4   Applications of Ensemble Learning.........................110
Review Questions...110

Chapter 6 Enhancement of Deep Learning Architectures................................112

6.1   Model Performance Improvement 112
6.2   Regularization 115
6.3   Augmentation...119
6.4   Normalization 120
6.5   Hyperparameter Tuning 123
6.6   Model Optimization 125
        6.6.1   Overview of Model Optimization 125
        6.6.2   Gradient-Based Optimization Algorithms...............127
        6.6.3   Other Optimization Algorithms...............................130
6.7   Neural Architecture Search (NAS) 132
        6.7.1   Overview of NAS 132
        6.7.2   NAS Process...133
        6.7.3   Search Space..134
        6.7.4   Search Strategies of NAS 136
        6.7.5   Strategies for Performance Measures.......................139
6.8   Adversarial Training 140
        6.8.1   Overview of Adversarial Training 140
        6.8.2   Types of Adversarial Attacks...................................141
        6.8.3   Adversarial Attack Generation Techniques 142
        6.8.4   Adversarial Attack Defensive Methods....................144
        6.8.5   Best Practices to Avoid Adversarial Attacks 145
Review Questions...145

Chapter 7   Performance Evaluation Techniques 147

        7.1   Overview of Performance Measures 147
        7.2   Types of Performance Metrics 148
               7.2.1   Confusion Matrix 148
               7.2.2   Accuracy 148
               7.2.3   Precision and Recall 150
               7.2.4   F-Measure 151
               7.2.5   Specificity and Sensitivity 152
               7.2.6   Receiving Operating Characteristic
                        Curve (ROC) 152
               7.2.7   Area Under the ROC Curve (AUROC)
                        and AUC 153
               7.2.8   Cross-Validation 153
               7.2.9   Kappa Score 157
               7.2.10   Grad-CAM Heat Map 157
               7.2.11   Metrics for Imbalanced Datasets 158
               7.2.12   Metrics for Regression Problems 159
               7.2.13   Summary of Performance Metrics 163
        Review Questions 163

Appendix – Frequently Asked Questions 165

References 173

Index 181

Preface

The rapid development of digital technologies has resulted in an explosive growth of data. Data engineering plays an essential role in many fields, including finance, medical informatics, and social sciences. This has led to increasing demand for career opportunities with the knowledge and experience of data science, with competence in computer programming. However, still, there is a global shortage of workforce whose skills span these areas.

Deep learning (DL) is an important and evolving area in data science that includes statistics and predictive modeling. It is concerned with algorithms inspired by the brain's structure and functions known as artificial neural networks. DL can automatically learn features in data, by updating learned weights at each layer. This book provides adequate theoretical coverage of DL techniques and applications. This book will teach deep learning concepts from scratch. We aim to make DL approachable by teaching the concepts and theories behind DL models. Thus, practitioners can grab the critical thinking skills required to formulate problems, design and develop models to make accurate predictions and support the decision-making process. Many academic institutions have embarked on DL education and research at various levels. At present, DL has become a forward-looking academic discipline with a wide range of real-world applications.

DL is extremely beneficial to data scientists in collecting, analyzing, and interpreting large amounts of data with efficient processing. There are many advantages associated with deep learning. For instance, DL techniques may produce new features from a small collection of features in the training dataset without any further human interaction. The ability to process large numbers of features makes DL techniques very powerful when dealing with unstructured data namely texts, images, and voices. More reliable and concise analysis results can be obtained as the prediction process is based on historical data. In the long run, it also supports improving prediction accuracy by learning from flaws. Although DL techniques can be expensive to train, once trained, it is cost-effective. Moreover, these techniques are scalable, as they can analyze large volumes of data and execute numerous calculations in a cost- and time-effective way.

DL applications are widely used in several industries like finance, transport, healthcare, automanufacturing, and advertising. For instance, DL is reshaping and enhancing the living environments by delivering new possibilities to improve people's life. For instance, in the healthcare domain, it helps in the early detection of cancer cells and tumors, improves the time-consuming process of synthesizing new drugs, and invents sophisticated medical instruments. Deep learning is used in the entertainment industry such as Netflix, Amazon, and Film Making. Netflix and Amazon use recommender systems to provide a personalized experience to their viewers using their show preferences, time of access, and history. Voice and audio recognition technology can be used to train a deep learning network to produce music compositions. Google's Wavenet and Baidu's Deep Speech can train a computer to learn the patterns and the statistics that are unique to the music. It can then generate a completely new composition. Additionally, the 'LipNet', which is developed by Oxford and Google scientists, could read people's lips with 93% success. This can be used to add sounds to silent movies. Further, in advertising, DL allows optimizing a user's experience. Deep learning helps publishers and advertisers to increase the relevance of the ads and boost the advertising campaigns.

Acknowledgements

We are grateful to all who helped improve the content and offered valuable feedback. Specifically, we thank K. T. S. De Silva and S. Dayarathna and T. Shyamalee for their contributions to collecting materials and designing graphics.

Abbreviations

Adaptive delta	**(AdaDelta)**
Adaptive gradient	**(Adagrad)**
Adaptive moment estimation	**(ADAM)**
Area under the curve	**(AUC)**
Area under the ROC curve	**(AUROC)**
Artificial intelligence	**(AI)**
Artificial neural network	**(ANN)**
Bi-directional encoder representations from transformers	**(BIRT)**
Capsule network	**(CapsNet)**
Carlini & Wagner attack	**(C&W)**
Convolutional neural networks	**(CNN)**
Deep learning	**(DL)**
Deep neural network	**(DNN)**
Differentiable architecture search	**(DART)**
Directed acyclic graph	**(DAG)**
Efficient neural architecture search	**(ENAS)**
Exponential moving average	**(EMA)**
Facebook-Berkeley-Nets	**(FBNet)**
False negative	**(FN)**
False positive	**(FP)**
Fast and practical neural architecture search	**(FPNAS)**
Fast gradient sign method	**(FGSM)**
Federated learning	**(FL)**
Generative adversarial networks	**(GAN)**
Geometric Mean	**(G–mean)**
Gradient descent	**(GD)**
Gradient-weighted class activation mapping	**(Grad-CAM)**
Internet-of-things	**(IoT)**
Jacobian-based saliency map attack	**(JSMA)**
Limited-memory Broyden-Fletcher-Goldfarb-Shanno	**(L-BFGS)**
Logarithmic loss	**(Log loss)**
Long short-term memory networks	**(LSTM)**
Machine learning	**(ML)**
Matthew's correlation coefficient	**(MCC)**
Mean absolute error	**(MAE)**
Mean squared error	**(MSE)**
Multilayer perceptron	**(MLP)**
Neural architecture optimization	**(NAO)**
Neural architecture search	**(NAS)**
Natural language processing	**(NLP)**

Peer-to-peer	**(P2P)**
Principal component analysis	**(PAC)**
Receiver operating characteristics	**(ROC)**
Rectified linear	**(ReLU)**
Recurrent neural networks	**(RNN)**
Region of interest	**(ROI)**
Residual network	**(ResNet)**
Root mean square error	**(RMSE)**
Root mean square propagation	**(RMSprop)**
Stochastic gradient descent	**(SGD)**
Stochastic neural architecture search	**(SNAS)**
True negative	**(TN)**
True positive	**(TP)**
Vision transformer	**(ViT)**
Visual geometric group	**(VGG)**
Youden's index	**(YI)**

1 Introduction

1.1 DATA-DRIVEN DECISION-MAKING AND SOCIETY

In a world where data-driven decision-making is becoming increasingly common, machine learning and artificial intelligence have come to be seen as valuable resources for making better and faster decisions. The advanced technological developments in the field of deep learning, which is a specialization of machine learning, have produced powerful tools for a wide range of applications. Advances in these fields have enabled computers to extract features, detect patterns, and make predictions about data and outcomes, with possible explanations to increase the trustworthiness of the applications. As a result, these techniques are being increasingly used in a wide variety of fields, including healthcare, education, finance, and social.

With the development of devices that generate large piles of data, we encountered a new concept of big data in the past decade. New analytic methodologies and data collection platforms have been created based on this concept and today we are exposed to a massive amount of data in every possible area of interest. The moment you are reading this, there are thousands of internet-of-things (IoT) devices, your mobile phones, or any other ubiquitous device that generates and sends data throughout the world among different networks.

Let us move our topic of discussion into data-driven decision-making, which is a common phenomenon found in the technical field, which provides new avenues to explore the usages in the decision-making process applied to different scenarios in various domains. For example, if we dive into the decision-making process powered by data in the business world, multiple causes can be used to describe the initiative. First, the collection of survey responses by the stakeholders of the business can be used to identify the enhancements or properties of their products, services, or features of their customers. Furthermore, with the use of advanced analytics, predictions can be made on how they are going to develop these new features to adhere to customers' liking. Also, some tests can be used as user testing to monitor their customers in using their products or services and enable them to identify potential issues associated with the development before the official release, which will lead to enhanced customer satisfaction and lesser defects or bugs.

From another point of view, when launching a brand-new product or to the market, data-driven decision-making enables us to analyze the patterns and behaviors in that

DOI: 10.1201/9781003390824-1

market and understand how the new product is going to perform in that particular market. This enables increased efficiency in market research and can immensely support many departments, including marketing, upper management, and even the development level to get clarity in the decision-making process. Most importantly, analyzing the patterns and shifts or new tendencies in the market based on demographic data applies to all sorts of businesses and different industries. Therefore, the data-driven decision-making process immensely supports organizations to determine opportunities or potential threats, where the organization can prepare with potential ways to tackle them.

In the applications of such data-driven decision-making, artificial intelligence (AI) takes a significant role. Since most of the activities are carried out by computers, the decision-making process is also supported by computational implementations more efficiently and effectively than human intervention. Developing artificial intelligence-enabled applications has been a topic of interest for a couple of decades. From email spam filters to autonomous cars, AI supports a wide range of applications, which are used to guide the human thinking process. In the early stages of artificial intelligence, knowledge was used to solve problems that were difficult to solve with human intelligence.

The usage of artificial intelligence to assist in decision-making will depend on a collection of factors including the current issues, vision, goals, nature of the application, and the type and quality of data to which it is exposed. This has become an essential tool to assist in making smarter and more impactful decisions. Therefore, decision-making is hugely benefited from data which is powered by the usage of advanced AI methodologies including machine learning and deep learning.

1.2   OVERVIEW OF DEEP LEARNING

Have you ever wondered how deep learning evolved from machine learning? This is understandable by comparing the differences between machine learning and deep learning. The improvements in computational technologies try to simulate the human intelligence process using machines. Initially, let us understand the concepts behind artificial intelligence (AI), machine learning (ML), and deep learning (DL) as shown in Figure 1.1.

- Artificial intelligence (AI) has the ability to perform tasks using machines that normally require human intelligence. It can be considered as a smart application that simulates the behavioral patterns of humans and learns without human intervention such as self-driving cars.
- Machine learning (ML) can be considered as a specialization of AI that consists of a stack of tools to analyze and visualize data, and predictions. It can learn using data without being explicitly programmed with a set of rules. This approach is based on training a model from datasets.
- Deep learning (DL) is a type of ML that simulates the human brain. It uses multiple layers in a deep neural network to progressively extract high-level features from the raw input. Their ability to analyze more complex relationships makes them particularly useful for modeling a wide variety of real-world problems.

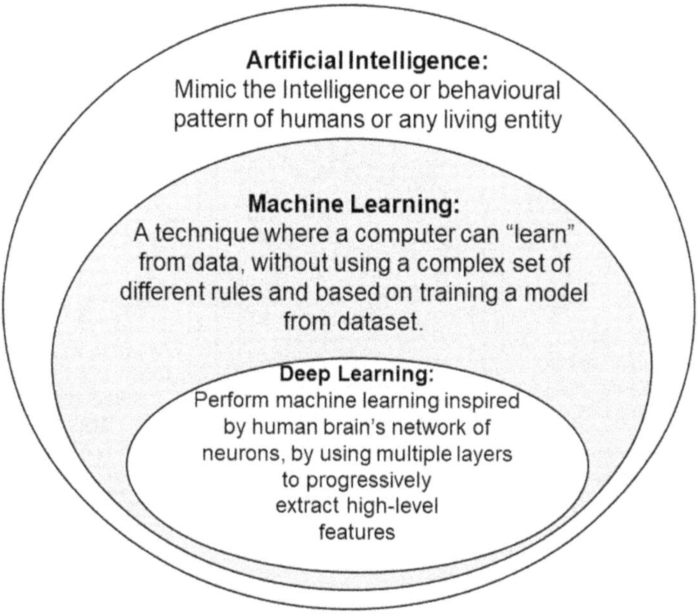

FIGURE 1.1 Overview of AI, ML, and DL.

Deep learning can be considered an evolution of machine learning. Earlier, machine learning algorithms were used to develop models for different applications. Many machine learning algorithms were developed to learn data and improved over time and are still used for making intelligent decisions. Deep learning is a subfield of machine learning with a multi-layered neural network that can learn and make intelligent decisions on its own. Deep learning models can learn the high-level features from the data on their own, while machine learning models need manually engineered features that are identified by domain experts. The evolution of these technologies is shown in Figure 1.2. Although the concept of deep learning was theorized earlier, it became more popular among data scientists recently. The main reasons can be stated as the recent advent of big data and high computational power GPUs. Most of the computationally infeasible algorithms and models became feasible technologically and concept-wise with the availability of a large amount of data, inexpensive data storage, and computation power. Consequently, deep learning models are widely used to solve real-world problems involving tasks like image recognition, speech recognition, and natural language processing.

Deep learning is defined as a family of machine learning models that are characterized by their deepness and generality. These models can learn complex, non-linear relationships between input data elements and the corresponding output. The implementation of these models uses artificial neural networks with multiple hidden layers; hence defined as deep neural networks (DNNs). While the data is transformed through these multiple hidden layers, each level learns the input data and transforms it into a slightly abstract and composite representation and eventually captures

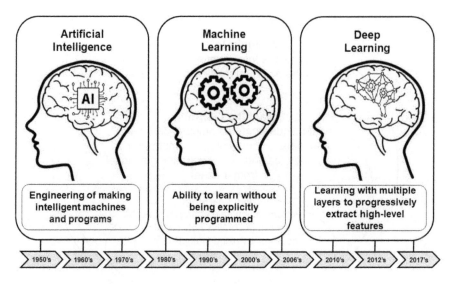

FIGURE 1.2 Evolution of technologies.

complicated relationships. Consider the image recognition application shown in Figure 1.3. For instance, the input can be a set of pixels in a matrix. The first layer may extract the pixels related to the edges. The second layer may compose the set of edges. The third layer may arrange the edges into different shapes. The fourth layer may predict the image. Likewise, the deep learning process extracts the features that can be learned at different levels. During this process, parameter tuning is needed to obtain optimal results by changing the number of layers and the size of the layers. Therefore, deep learning learns progressively to extract features and make optimal predictions.

Overall, in a DNN data flows in a feed-forward direction from the input to the output layer. A neural network consists of mainly three types of layers namely, input, hidden, and output layers. Initially, the DNN creates the linkages between neurons and assigns random numerical weights for the connections. In this process, the weight values and inputs are multiplied and generate an output within the range 0 and 1. The algorithm adjusts the weights until the expected prediction accuracy is reached. Subsequently, the algorithm makes several parameters more prominent in hidden layers, until the optimal equations are obtained to process all the data.

Considering the data transformation between inputs and outputs, machine learning needs the necessary representation of the data that is suitable for algorithms to transform into output. However, deep learning models learn many layers of transformation while each layer offers a representation at one level. For example, layers that are near the input contain fewer details of the data while layers that are near the output have high-level data representation with the concepts used for the data discrimination. The deep learning models can be identified as multi-level representation learning. These models with many layers are more capable of extracting low-level perceptual data than other models. This not only performs better than other shallow models, but it

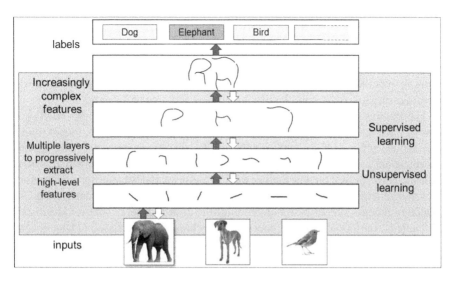

FIGURE 1.3 Overall process within a deep learning model.

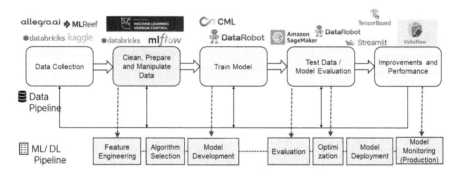

FIGURE 1.4 Life cycle of a deep learning process.

also has an accurate and automated feature engineering process by eliminating many boundaries in areas such as natural language processing, computer vision, and speech recognition.

With the overview of the deep learning concept, it is important to identify the data processing stages to produce insights and predictions to obtain the outcome in practice. As shown in Figure 1.4, the life cycle of a deep learning process mainly consists of data acquisition, preprocessing, training, testing, evaluation, deployment, and monitoring. We learnt that the overall effectiveness of the model depends mainly on the data. Generally, data can be private or public and collected using surveys or experiments. With the availability of data, a data scientist needs to understand the data by exploring the structure, relevance, type, and suitability of data. Although data preparation is a time-consuming task, it plays an important role in the life cycle, to derive new features from the existing data. This, exploratory data analysis is required to identify the affecting factors using the data distribution between different feature variables before the actual model design. Accordingly, in the data flow pipeline, the data should be preprocessed to clean, remove outliers, manage missing data, normalize, and augment before feeding into the training model. Considering the deep learning pipeline, the feature engineering techniques are aligned with the data preprocessing to extract and identify informative features of the data.

The training algorithm can be selected considering different factors such as the type of data, nature of the application, and resource availability. The model training and testing processes are engaged in tuning the hyperparameter and applying optimizations to generate better results. These models should be designed to learn the data and perform well on new data as well, by ensuring the balance between performance and generalizability. Once the model is evaluated by testing on unseen data. The modeling process should be reiterated until the desired level of metrics is achieved. A detailed description of these concepts and techniques is discussed in later chapters of this book. Once the final model is deployed, the application is monitored for further improvements. In practice, several frameworks and technologies are available to ease the processes in the deep learning life cycle. Furthermore, different tools and frameworks are utilized to accomplish these processes.

Deep learning is still evolving with novel ideas of big data processing with artificial intelligence. Therefore, you need to better understand deep learning techniques and their key concepts for the development of innovative applications. This book will provide you with the theoretical background on basic deep learning techniques, neural networks, deep learning models, types of deep learning approaches, architectural enhancements, and evaluation techniques.

1.3   BIAS AND VARIANCE

In general, a machine learning algorithm aims to correctly determine the mapping function to predict a variable y (output) given x (input). However, there is always a difference between model predictions and actual results, which is known as prediction error. Therefore, it is required to have an awareness of the bias and the variance errors, when training a machine learning model. These fundamental concepts on parameter estimation, bias, and variance are useful in identifying model characteristics on generalization, overfitting, underfitting, and accurate predictions.

Let us start with the basics of point estimation and interval estimation with simple statistics. Point estimation calculates a single value of an unknown parameter such as a model weight or a whole function. Since it estimates the relationship between the input and output, it is also known as a function estimator. For example, the sample mean is considered as a point estimation. The interval estimation results in a range of values that a parameter can remain. For example, the confidence interval is considered as an interval estimation.

Let us denote the point estimation of a parameter θ by $\hat{\theta}$ to differentiate the estimation of parameters from their actual values. Let $\{X1, X2, .., Xm\}$ be an independent and distributed set of data points. The point estimator can be defined as a function of data, $\hat{\theta} = f\{X1, X2, .., Xm\}$, where '$f$' returns a value close to the actual value of θ.

In general, a function, where the predicted output is closer to the actual value of θ, is considered as a good estimator. It is important to review the properties of these estimators. The following sections describe the bias and variance. A correct balance between bias and variance is important to generate accurate predictions.

1.3.1 SKEWNESS OF DATA

Skewness determines how far a random variable's probability distribution deviates from the normal distribution (probability distribution without any skewness), as shown in Figure 1.5. Also, skewness indicates the direction of outliers.

For example, let us consider the case of positive skew data, where a large number of data instances consist of small values. This results in better training performance at predicting data instances with lower values. Thus, there is a 'bias' towards lower values, in this scenario. Considering the direction in this case, most of the outliers appear on the right side of the distribution. Thus, there is a variance in data.

1.3.2 BIAS

The term bias can be defined as the deviation between the predicted value by the deep learning model and the actual output or the ground truth. When the bias is a higher value, it indicates a large error in the output of the model. Also, it can indicate the imbalance of the dataset. Generally, we expect a model to have a low bias to prevent issues such as the underfitting of data to the model. This can be explained as a systematic error of the training model, which skews the result in favor or against the actual output. Bias shows the matching of the dataset to the model as follows:

In a high-bias situation, the dataset does not match the model.
In a low-bias situation, the dataset fits with the training model.

The following indicators help to identify a high-bias model:

Failure to capture the data trends
Potential towards underfitting
More generalized/overly simplified
High error rate

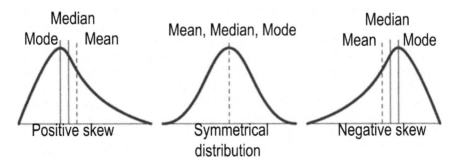

FIGURE 1.5 Data distributions.

The bias of an estimator can be defined as in (1.1), where $\hat{\theta}$ denotes the estimation of a parameter where the actual value is θ, and $\hat{\theta}m$ is the point estimator. The term $E(\hat{\theta})$ is an expectation of the data. If Bias $(\hat{\theta}) = 0$, the estimator of $\hat{\theta}$ is considered as unbiased, as $E(\hat{\theta})$ is equal to θ.

$$\text{Bias } (\hat{\theta}) = E\ (\hat{\theta}) - \theta \tag{1.1}$$

1.3.3 VARIANCE

Variance indicates the expected difference between the observed data instances from the average value. Thus, variance indicates the data spreading within the sample set. Both bias and variance of an estimator are calculated for a dataset. This variance measures how the estimate would vary as the dataset is changed independently from the underlying data generating process. In other words, it indicates the changes in the model with different parts of the training data set.

Since the target function is estimated from the dataset, it is acceptable to have some variance. However, it should not vary drastically from one dataset to another, which indicates that the estimator is good at understanding the hidden mapping between inputs and outputs. It can be considered as an indicator of the uncertainty in the data. A high variance indicates that the estimator does not generalize on unseen training datasets. In that case, the model shows high performances on the training set but gives high error rates on the testing set. It is good to have a relatively low variance for an estimator.

The following indicators help to identify a high-variance model:

Noise in the dataset
Potential towards overfitting
Complex models
Trying to include all data points closer.

1.3.4 TRADE-OFF BETWEEN BIAS AND VARIANCE

As we already discussed, bias and variance are used to show the errors in an estimator. Overall, bias and variance calculate the difference from the actual value, and the deviation from the expected estimator with the changes in the dataset, respectively, during model training and testing. It is expected to have a balance between these terms, that is, low bias and low variance. Figure 1.6 visually explains the bias and variance in feature classification of 2D space. A detailed description is included in Chapter 2. Additionally, Table 1.1 states a comparison of bias and variance, and Table 1.2 shows the trade-off between bias and variance with different training and testing error values.

Accordingly, if the model has a high bias, it has made more assumptions about the target function. Underfitting may result from missing significant relationships between characteristics and outputs. The changes in training data will produce substantially diverse target functions if a model has a high variance. As a result, the model overfits and starts to learn the random noise instead of the output. Since the model has a greater capacity to learn from the training data, increasing model complexity would

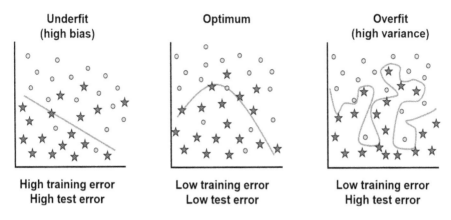

FIGURE 1.6 Bias and variance of a classifier.

TABLE 1.1
Comparison of Bias and Variance

	Bias	Variance
High model complexity	Low bias	High variance
Causes	High bias results in underfitting	High variance results in overfitting.
Feature	Low bias indicates fewer target function assumptions are made.	Low variance means that similar target functions would result from training data changes.

TABLE 1.2
Trade-off Between Bias and Variance

Train error	Very low	Relatively high	Relatively high	Very low
Test error	High	Relatively high	Very high	Low
Bias-variance	High variance	High bias	High bias and high variance	Low bias and low variance

typically result in a decrease in bias error. The variance error will rise as a result, though, and the model may start to pick up on noise in the training set.

However, since bias and variance are inversely connected, it is hard to have a model with a low bias and a low variance. Therefore, the trade-off between these terms can be stated as follows.

High-bias models will have low variance.
High-variance models will have a low bias.

Generally, when the model is simple with few parameters then it may have high bias and low variance. Here, the model may not have the risk of generating inaccurate results, but it will not match the dataset. In contrast, if the model has many parameters, then it will have high variance and low bias. In this situation, although the model fits with the dataset, there are more chances to predict inaccurate results. These aspects indicate the model's flexibility to obtain an optimal model. For example, if the model does not fit with the dataset, it will have a high bias. This leads to an inflexible model with low variance.

This trade-off can be addressed, and the most suitable model can be selected by comparing the mean squared error (MSE) of the estimators as in (1.2). The estimators with less MSE can keep both their bias and variance in an acceptable range. It should be noted that these biases and variances are linked with capacity, overfitting, and underfitting concepts in deep learning.

$$\text{MSE} = E\left[(\hat{\theta} - \theta)^2\right] = \text{Bias }(\hat{\theta})^2 + \text{Var }(\hat{\theta}) \tag{1.2}$$

Consider, Figure 1.7 with model capacity against the error. The capacity of the model indicates its capability to suit a variety of functions. When the capacity increases, the bias of the model tends to decrease, and variance gets increased. This produces another U-shaped curve, which represents the generalization error. As capacity varies, there is an optimal point in the graph that denotes a good balance between bias and variance that minimizes the error. However, a learning algorithm can handle some variance. A model that is optimally balanced between bias and variance is neither overfitting nor underfitting.

A trained model with the lowest bias versus variance trade-off for a specific dataset is the desired outcome. Techniques such as cross-validation, regularization, dimension reduction, stop training early and use mode data will help overcome bias and

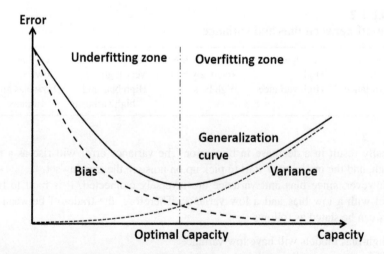

FIGURE 1.7 Model error variation with respect to model capacity.

variance errors. The following tasks can be applied to address the trade-off between bias and variance.

- Increase the complexity of the model. This decreases the overall bias while increasing the variance to an acceptable level. This aligns the model with the training dataset without incurring significant variance errors.
- Increase the training dataset. This is the preferred method when dealing with overfitting models. This allows users to increase the complexity without variance errors that negatively impact the model with a large dataset. A learning algorithm can be generalized easily when there are many data points. However, when the data is underfitting or the model shows low bias, the model is not sensitive to the dataset, even in a large dataset. Therefore, for models with high bias and high variance, using a large dataset is a feasible solution.

1.4 SUPERVISED AND UNSUPERVISED LEARNING

The real-life problems can be categorized into two main sets, that is, yes or no questions and quantification questions. The yes or no question often requires binary decision-making procedures, where the evidence or data is presented. The quantification problems require data and calculations to find the quantified answers. The same approach is taken in using computers to answer these questions. However, as we see in real-life problem solving, these two types of questions follow two distinctions in the methodology using computers. The problem-solving process with yes or no answers is known as classification and the other type of problem-solving is known as regression to predict a quantified amount. Therefore, these two approaches follow different sets of machine learning algorithms to answer the questions we address. The classification problems can be further divided into supervised and unsupervised learning.

Supervised learning predicts the label of a given data item, given all the resources to navigate through the problem. Here, the dataset needs to be annotated into the expected outcome or classes of the model. If we incorporate supervised learning into the actual learning process, imagine that you are given all the problems, and sample questions and the teacher is teaching you the things that you need to follow to understand the problem and design a solution. In machine learning terminology, for each data point, the expected answer is named as the label. The sample data associated with the label are called features. The goal of supervised learning is to produce a mapping function that maps the features or input data into the label. The mapping function can differ based on the problem. Let us consider a real-life example to understand this terminology. Consider a situation where a doctor needs to predict whether a patient is going to have a stroke after examining and based on his past patient history. In this situation, the patient's history may contain information about previous strokes and their causes, which are the labeled examples provided. The current observation is the input data and the outcome should be whether the person is going to have a stroke or not. Here, the doctor is the decision-maker. However, in a machine learning context, the computer provides this decision using the provided labeled data.

Some of the real-world examples that use supervised learning are listed as follows.

- Predict cancer vs non-cancer given computer tomography images.
- Identify fraud and non-fraud signatures in financial documents.
- Predict the stock prices for the next month based on this month's financial data.
- Identify spam and non-spam emails.
- Classification of positive and negative sentiments from tweets.

In contrast, unsupervised learning categorizes unlabeled datasets, by identifying the hidden patterns and extracting useful information without human intervention. Here, the number of questions and answers solely depends on the quality and the hidden information in the dataset. Let us understand this concept using a real-world example. Consider a supermarket that wants to group its customers, to recommend products. They can apply a data-driven approach, by grouping the customers based on their age group and then deriving insights from these groups.

Some of the examples of unsupervised learning are listed as follows.

- Group a set of random photos into landscape photos, pictures of dogs and cats, babies and mountain peaks, etc. This is known as clustering in machine learning terminology.
- Find a group of small numbers of parameters that can be used to explain the data. This extracts the most important features from the dataset, which explain the dataset the best. This procedure is known as principal component analysis in machine learning terminology.
- Identify the functional proteins that affect the most in cancer diagnosis.
- Identify the patterns in financial fraud activities.
- Identify the important minimum set of dimensions in magnetic resonance imaging data.

1.5 SUPPORTIVE TOOLS AND LIBRARIES

A complete platform to execute deep learning models can be created using a set of programming languages, machine learning libraries, services and web applications. Figure 1.8 shows a set of tool stack that supports building the deep learning models efficiently in terms of resource utilization, maintenance, team efforts, and user experience. We discuss some of the platforms that support deep learning.

FIGURE 1.8   Tool stack to support the deep learning process.

1.5.1 TENSORFLOW

TensorFlow is an open-source platform for fine-tuning large-scale machine learning applications using different libraries, tools, and resources. Initially, this framework was developed by Google for their internal usage and later provided as an end-to-end machine learning platform in the public domain. Among several functionalities, it mainly supports model training and inference of deep neural networks. Since there are a large amount of data to process using complex algorithms, it is required to store data compactly and feed it to the neural network. Tensors provide a better way to represent data as an *n*-dimensional matrix or a vector. Since these tensors hold data in different known shapes, the shape of the data can be identified with the dimension of the matrix. After storing data in tensors, the relevant computations can be performed in the form of a graph. These tensors can be derived either from input data or as computation results of an operation that conducts inside the graph. The input data goes into the graph at one end and then flows through various operations and comes out at another end as an output. All these operations in the graph are known as graph op nodes that are connected by tensors as edges. These graph frameworks can run on multiple CPUs, GPUs, or mobile operating systems. Accordingly, some of the benefits of using TensorFlow can be stated as open-source, platform independence, train on CPU and GPU, high flexibility, autodifferentiation and manage threads and asynchronous computation.

1.5.2 KERAS

Keras is an open-source deep learning API written in python that executes on machine learning platforms. This provides an interface to solve complex learning problems aiming at deep learning techniques. This API acts as a high-level wrapper to create deep learning models, define their layers, and compile models. However, this does not support other low-level API such as generating computational graphs and making tensors and sessions. Keras supports multiple backends for the computation such as TensorFlow, Theano, CNTK, and PlaidML. TensorFlow uses Keras as its official high-level API, supporting many in-built modules to compute neural networks.

Keras offers a simple API that reduces the complexity of making neural network models and allows you to implement the codes with a simple set of functions. Since Keras supports multiple cross-platforms, a given backend can be selected depending on the requirements. When using TensorFlow with Keras API, we can easily create customized workflows based on the requirements. Also, this is much easier to learn as it provides a python frontend with a high level of abstraction. Keras can be deployed on devices like iOS, Android, Raspberry Pi, Cloud Engines, or Web Browsers with .js support. Also, Keras runs on both GPU and CPU, with the support of in-built data parallelism to process large data volumes for model training. Therefore, this can be used easily and efficiently as a high-end API to create deep learning networks.

Let us learn the main steps in creating a simple Keras model. The basic elements of Keras are models and layers. Initially, we need to define a network by adding layers to support data flow in the selected model type. There are two types of models namely sequential and functional. Then we need to define the loss function, optimizer, and the other matrices to calculate the model accuracy, and compile the model to convert it into a machine-understandable format. Next, the model can train, evaluate, and predict the results. Additionally, the Keras functional API can be used to build arbitrary graphs of layers or develop models in complex architectures.

1.5.3 PyTorch

PyTorch is a python-based open-source machine learning framework. It uses an optimized tensor library for deep learning using GPUs and CPUs. One of the main high-level features of PyTorch is its dynamic computational graph based on automatic differentiation. In contrast to TensorFlow, where we need to first define the entire computational graph before running the model, PyTorch allows us to define the graph dynamically. The PyTorch library is designed for more efficient use by tracking the model built in real-time. Since the developers can dynamically change the behavior of the graph, it is easier to use than TensorFlow. Also, PyTorch enables GPU-accelerated tensor computations and effective data parallelism. However, compared to TensorFlow, PyTorch provides limited visualizations during the training process.

1.5.4 Jupyter Notebook

Jupyter Notebook is an open-source web application. Developers use this environment to create and share documents with source codes, text, and visualizations. This helps to perform end-to-end workflows in data science such as data preprocessing, model building, model training, data visualization, and many other related works. Jupyter notebooks can use to write codes in independent cells and execute them individually. Therefore, this allows testing specific blocks of code without executing the entire script of code as in many other IDEs. This is a flexible and interactive platform that is widely used in data science.

1.5.5 NumPy and Pandas

The libraries NumPy (Numerical Python) and Pandas (Panel Data) are important in deep learning due to their matrix computation capabilities. Both are open-source Python libraries. NumPy consists of multi-dimensional array objects and a set of procedures to process them for numerical computations. Thus, it supports processing large matrixes using mathematical functions. Pandas is built on top of the NumPy package and supports functionalities such as loading, manipulating, preparing, modeling, and evaluating tasks for multi-dimensional data. NumPy and Pandas modules are best suited for numerical and tabular data, respectively.

1.5.6 Tensor Hub

TensorFlow hub provides a repository of pretrained models as off-the-shelf models to be used in machine learning tasks. These models are used for fine-tuning to build real-world applications and learning purposes with few lines of code. The models in this repository support a variety of applications, such as pattern recognition, object detection, audio processing, and natural language processing.

REVIEW QUESTIONS

1. What are the advantages of using deep learning based applications in the real world?
2. Explain the importance of balancing bias and variance.
3. What are the problems of having high-dimensional data and explain possible approaches to address those issues?
4. What aspects need to be considered when selecting the correct support tool or support library when solving a learning problem?

2 Concepts and Terminology

2.1 UNDERSTANDING NEURAL NETWORKS

The idea of neural networks comes from the neuroscience understanding of the support of the human brain for memory and decision-making. The work on mammalian vision systems by David Hubel and Torsten Wicsel paved the way for an understanding of biological neural networks. They explored the communication between the brain cells that enables us to see the world. If we go deeper into the biological level, the special cell type named neuron is responsible for consuming data and generating responses accordingly. Figure 2.1 shows the architecture of a biological neuron. The dendrites are the elements that capture the input signals into the system and the output signal traverses through axons and outputs through the synapse to other neurons. Here, spikes are short electric pulses that are handled and generated through this system. Some of these signals become significant by triggering some other neurons to fire, making the connections stronger or weaker.

Consider a linear activation function $f(x)$. This process can be reproduced in a computational system by implementing a function that gets the weighted inputs and outputs the sum of weighted inputs with a bias as shown in Figure 2.2. In a non-linear function, there is a non-linear relationship between the input and output. One key difference with the biological system is that this simplified computational method does not imitate the establishment or destruction of the neurons and ignores signal timing. The enormous number of possibilities of this discovery of computational neural networks with the analogy with biological neurons generated a hype to develop applications that were earlier thought impossible.

The first implementation of artificial neural networks (ANNs) consisted of a single layer of perceptron, but the design could not solve non-linear classifications. This problem was addressed by introducing multiple perceptron layers, changing their weights randomly and trying to figure out the mappings from inputs to outputs. This approach has made artificial neural networks stagnate for a long period. Over the years, the solution was presented by predicting a continuous output rather than using a binary output value. Although it does not seem like a massive development, it was the key change made in ANNs to perform with increased accuracy and performance. This change was able to depict the underlying mathematical base as separate regions that operates effectively and model them as partial derivations.

DOI: 10.1201/9781003390824-2

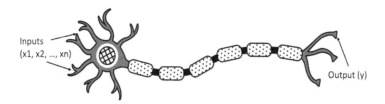

FIGURE 2.1 Biological neuron architecture.

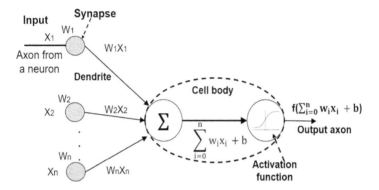

FIGURE 2.2 Descriptive neural function.

The ImageNet challenge was the next key milestone in the development of artificial neural networks. Deep neural network (DNN) architectures were able to learn new features and carry out classification tasks. In 2012, the model AlexNet won the ImageNet challenge, paving the way for neural architectures. Since then, several neural architectures have been developed to address different kinds of problems, including image classification, video classification, feature identification, and image reconstruction.

If we dive into the mathematical approach in ANNs, we learn about the important theory of logistic regression, which is considered as the basic theoretical approach in ANNs. In general, regression is a set of methods to model relationships between independent variables and dependent variables. In real-world problems using machine learning, the independent variables are known as inputs and the dependent variable is known as the output. Therefore, it is required to find a mapping between inputs and outputs known as predictions.

Consider an online shopping website that suggests items to buy based on the customers' details and their buying patterns. A model to recommend items can be developed by using a dataset consisting of many variables such as customers' historical buying items, personal data involving age, sex, location, and user data with views and likes. In deep learning terminology, the dataset is divided into the training set and test set. The data corresponding to one record is known as a data instance. The model is trained to predict an item to buy, and it is called as the label of the target object. All the independent variables considered to produce the prediction are known as features.

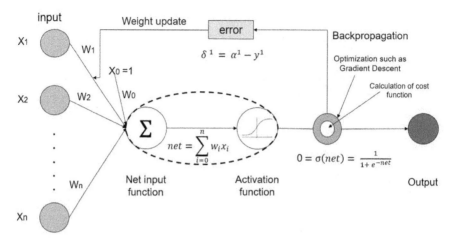

FIGURE 2.3   Overview of terminologies.

Let us consider an input denoted as x_i, which is a combination of input features of $[x^i_1, x^i_2 ..]$, and the corresponding label denoted as y_i. The linear equation of the target can be presented as in (2.1), where w_1, w_2 are the weights, and b is the bias. The weights are essential to showcase the impact of the influence of the input features, where bias indicates the prediction value when all inputs are zero. The bias variable is used to limit the expressivity of the model. Our task is to find the best weight vector, which minimizes the deviation between the predicted and the actual value. Consequently, this is followed by quality measures and a mechanism to improve the model quality.

$$Y_i = w_1 \cdot X_1 + w_2 \cdot X_2 + \ldots + + w_n \cdot X_n + b \qquad (2.1)$$

In summary, a neural network consists of connected neurons, which have their weight, bias, and activation function for the classification and regression processes. In a neural network, it is important to update the weights and the bias parameter. If they update correctly, the model will converge into the global minimum point easily. The error is defined as the difference between the actual and predicted output. Figure 2.3 depicts an overview of terminology associated with a simple weight-updating process of a neural network.

2.2   REGRESSION

Although this book does not focus on regression techniques in detail, this section gives an overview of regression for better understandability. Regression analysis is a method that identifies the relationship between the independent variables and dependent variable in the dataset. Based on the type of relationship linear or non-linear regression techniques will differ and the target variable is always a continuous value in regression. This is mainly used to forecast the trend, and identify the strength of the predictor and time series. Different regression techniques, such as linear regression and logistic regression, are used based on the nature of the data.

2.2.1 LINEAR REGRESSION

Generally, linear regression learns the linear relationship between the features and the target. It is a supervised based learning algorithm used for the predictive analysis of continuous values, and therefore, cannot learn the complex non-linear relationship. For example, consider a set of data points. The linear regression finds a line that the data points fit well. Thus, the output of a new data point can be predicted in a way that fits along the best-fit line. Since it solves linear problems by predicting continuous dependent variables using one or many independent variables as shown in Figure 2.4.

Consider an input vector $x \in R^n$ and an output value $y \in R^n$. The output is a linear function of the input such that $y = mX + c + e$, where y, m, X, c, e denote target, gradient, predictor, bias, and error. We get the best fit line with the lowest prediction error by changing the m and c values. However, linear regression does not suit learning complex non-linear relationships. Additionally, since this is sensitive to outliers, this will work well for small-size datasets.

Let us learn the basic principles of linear regression using mathematical representations. Let the model predicted value is $\hat{y}$ and the actual output is y. We can define the output to be, $\hat{y} = w^T x$, where $w \in R^n$ is a vector of parameters. These parameters are a set of values that controls the behavior of the system. In this case, w is defined as a set of weights (w_i) that determines the effect of each feature (x_i) on the final prediction. If the feature x_i receives a positive weight w_i, that means, the increase of the feature x_i increases the value of the final prediction from the model $\hat{y}$. Similarly, if the features receive a negative weight, the increase of features causes a decrease in the value of prediction. If the weight is zero, it does not affect the final prediction. Linear regression is often used with

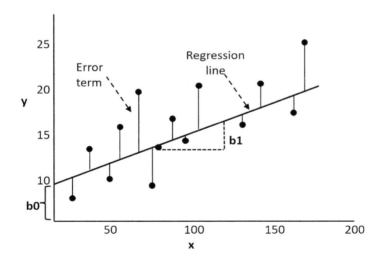

FIGURE 2.4 Linear regression problem.

an intercept term b, which is known as bias. This defines that the terminology is biased towards b in the absence of any input value. The basic principles of linear regression can be used to develop complicated and advanced learning algorithms.

2.2.2 LOGISTIC REGRESSION

We discussed linear regression, which predicts continuous valued quantities as a linear function of independent variables. However, it is not a good solution to predict binary-valued labels. This is addressed by logistic regression, by predicting the probability of a given example belonging to a particular class using the sigmoid function to represent probability distribution over binary value. Thus, a neural network without a hidden layer and a sigmoid activation in the neurons associated with the output layers is called logistic regression. This is a machine learning method that uses the concepts of probability in classification. This regression is used when the target variable is discrete that is 0 or 1, and the sigmoid function denotes the relationship between the predictor and the target variable as shown in Figure 2.5.

When logistic regression is used for multi-labeled classes, we use the Softmax activation function, which is known as Softmax regression or multinomial logistic regression. Thus, it is often used as the output of classifiers to represent probability distribution over multiple classes. Logistic regression is used when the dataset is large. Also, there should not be any correlation between the independent features.

2.2.3 OTHER REGRESSION METHODS

Several types of regression techniques such as lasso regression (L1), ridge regression (L2), polynomial regression and Bayesian linear regression are available. This section gives an overview of other regression methods. L1 and L2 are used to reduce

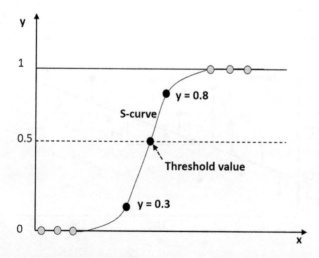

FIGURE 2.5 Logistic regression problem.

the overfitting in linear regression and reduce the least-squares error. These are used when there is more correlation between predictors and the target variables. The L1 and L2 regularization methods are further discussed in Section 2.5.5. Generally, L1 regularization is used when many feature sets provide sparse solutions. It performs both regularization and feature selection. When the independent variables are highly collinear, L1 regression gets only one variable and makes other variables shrink to zero, thus reducing overfitting.

In polynomial regression, the relationship between variables is given by the n-th degree. Although it tries to obtain the best-fit curved line that goes through all the data points, by reducing the mean squared error (MSE), this can cause overfitting where the model performs well only for the training set, without generalizing. Bayesian regression is another technique that uses the Bayesian theorem to identify the coefficients. Here, it finds the posterior distribution of features more stable than linear regression.

2.3 CLASSIFICATION

Now, let us see the difference between regression and classification. Classification is a supervised learning algorithm that performs predictions with labeled datasets. Although, it is used for the same tasks as in regression, the difference lies based on the applications. For instance, regression techniques are used to predict the continuous values such as temperature, item price, salary, and age; whereas classification classifies the discrete values such as dog or cat, health or unhealthy. Figure 2.6 and Figure 2.7 show visual representations of classification and regression.

Classification divides the dataset into different classes. As shown in Figure 2.7, A and B are two classes. There are similar features within the class and different features between classes. Accordingly, classification algorithms divide the dataset into different classes based on a set of parameters. Initially, a computer program is trained on a training dataset. Then based on the trained model, the test set is classified into different classes. Thus, it can be stated as a mapping function that maps the input to a discrete output. There are different types of classification, such as binary classification, multi-class classification, and multi-label classification. A variety of

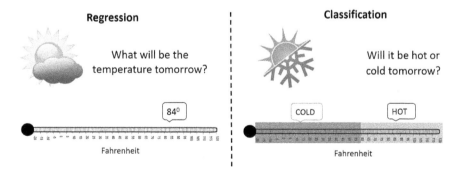

FIGURE 2.6   Example: classification vs regression.

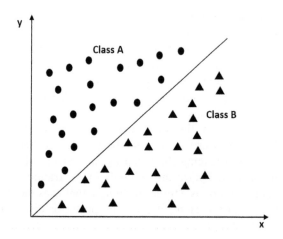

FIGURE 2.7   Classification task.

classification techniques are available such as convolutional neural networks (CNNs), recurrent neural networks (RNNs), generative adversarial networks (GANs), long short-term memory networks (LSTMs) and multi-layer perceptrons (MLPs), which we discuss in later chapters.

2.4   HYPERPARAMETERS

2.4.1   OVERVIEW

Hyperparameters are the variables that define the structure of the model network. These variables specify the training parameters of a model. The hyperparameters, such as the learning rate, are set before optimizing the weights and bias during model training. Some other hyperparameters are the number of hidden layers, batch size, number of epochs, and number of nodes in each layer, which are used for the training process. We can list the steps of hyperparameter tuning as follows. Additionally, Table 2.1 states the impact of different hypotheses.

  Step 1 — Decide the network structure.
  Step 2 — Adjust the learning rate.
  Step 3 — Select an optimizer and a loss function.
  Step 4 — Decide the batch size and number of epochs.
  Step 5 — Random restarts.

2.4.2   WEIGHT INITIALIZATION

In neural networks, the weights are initialized to small random numbers, in such a way that it will prevent the exploding or vanishing of the activation outputs during the feed forwarding in the network. If either of the above situations occurs, the loss gradients become very large or very small to flow backwards in backpropagation and

TABLE 2.1
States Different Types of Hyperparameters and Their Approximate Sensitivity

Hyperparameter	Approximate Sensitivity
Learning rate	High
Optimizer type	Low
Optimizer parameters	Low
Batch size	Low
Weight initialization	Medium
Loss function	High
Model depth	Medium
Layer size	High
Layer parameters (kernel size)	Medium
Weight of regularization	Medium
Non-linearity	Low

eventually the model may not converge. Here, in feedforward propagation, the input is used to compute the intermediate function in the hidden layer and used to get the output. During backpropagation, the model weights are altered repetitively to obtain the prediction output closer to the actual output.

These weights should not be the same and should allow different learning styles. If the weights are the same, then every neuron in each layer will learn the same features in the same way. Thus, the weights should have a good variance to learn new features. The selection of the weight initialization techniques is mainly based on the type of the dataset and the activation functions. The neural network uses matrix multiplication at its core operation. Matrix multiplication is mainly used in the prominent layers in the preliminary stages of the neural network. The matrix multiplication of layer inputs and weights is calculated as a resulting matrix and after applying the activation function, the resulting matrix is applied to the next layer.

We will discuss two weight initialization techniques: zero initialization and random initialization. Zero initialization defines the bias variable to be zero in the first step and assigns weights to zero. Here, the derivative concerning the loss function will be the same for every weight in the weight matrix. This makes the underlying hidden neurons become symmetric and ultimately performs worse than a linear model. Therefore, zero initialization does not result in successful classification. Random initialization is used in many computational problems, such as advanced searching algorithms like gradient descent. It is better than zero initialization. However, we cannot guarantee that the weights become too high or too low values, which can drive away from the classification or learning process.

If we initialize the weights with very high values, then the derivative terms of ($wx+ b$) become higher. When an activation function, such as sigmoid, is used, the function tends to map the values to nearly 1. This causes the gradient descent to make a slower progression in finding the minimum value, which results in increased learning time. On the other hand, if we initialize the weights as too low, it gets mapped to 0 in the activation function and results in a vanishing gradient problem. However, to utilize the randomness in the search process there is a need for stochastic optimization algorithms.

Therefore, the following aspects should be considered when initializing the weights of a neural network.

- The model efficiency: how long the training would take place.
- How to handle the vanishing or exploding gradient problem.

The widely used weight initialization mechanisms in neural networks can be described as follows.

1. He Initialization

Also known as Kaiming Initialization, and used to initialize weights when using non-linear activation functions, such as rectified linear (ReLU). These functions are discussed later in this chapter. This method calculates the weights as a random number with a Gaussian probability distribution (G) with a mean of 0 and a standard deviation of sqrt($2/n$), as shown in (2.2) to avoid vanishing or exploding the magnitudes of input. Here, n is the number of inputs to the node.

$$\text{weight} = G\ (0,\ \text{sqrt}(2/n)) \tag{2.2}$$

2. Xavier Initialization

This approach is also known as Glorot initialization and is used with neural networks that use sigmoid or tanh activation functions. Similar to He initialization, the weights initialization is based on normal or uniform distribution with a minimum 0 and standard deviation. The Xavier initialization method initializes the weights in a way that the activation variance is the same across each layer. The gradient exploding or vanishing can be prevented by having a constant variance. This method calculates the weight as a random number with a uniform probability distribution (U) between the range $-(1/\text{sqrt}(n))$ and $1/\text{sqrt}(n)$, as shown in (2.3), where n is the number of inputs to the node.

$$\text{weight} = U\ [-(1/\text{sqrt}(n)),\ 1/\text{sqrt}(n)] \tag{2.3}$$

2.4.3 ACTIVATION FUNCTION

The activation function is used to learn complex data patterns using a neural network by deciding the important features that pass to the next neuron while suppressing irrelevant data points. This behaves the same as the function of an activation function in the biological neural network. Similarly, the output from the preceding node is transformed into a format that can be passed as the input to the next neuron. This function can be defined in both linear and non-linear forms, where the non-linear form is widely used. For example, when there are no activation functions in a neural network, the neuron will perform only a linear conversion on the inputs using weights and biases, and all the hidden layers have similar behavior. Thus, when there is no activation function, it is hard to learn complex tasks and the model will behave as a linear regression model.

The activation functions are useful to keep the output value to be bounded by a threshold value, which acts as an upper limit. Here, the input to the activation function comes as a product of the weight and input plus the bias values, which are not normalized or have restrictions in their ranges. Therefore, to normalize the output to avoid unnecessary computations we can use an activation function.

As we have discussed earlier, one key drawback of the earlier neural networks was their inability to adjust non-linear inputs and outputs in the network. Hence, having an activation function enables non-linearity in the neural network. To highlight the importance of non-linearity in models, let us draw a real-world example of a classification problem. Imagine that you are given a problem identifying patterns of the dataset consisting of weight, blood pressure, and age and you are asked to find out the patterns of a smoker and non-smoker. This classification scenario is a non-linear problem and requires using an activation function.

There are multiple things that we need to consider when designing an activation function. The first one is to beware of the vanishing gradient problem. The neural networks learn through backpropagation using the gradient descent method to adjust weights to minimize the loss in each epoch. The activation function limits the output of the layer to 0 or 1 at the end. The network tends to backpropagate those values to map between the input and output and to assign the weights to minimize the loss. Thus, having 0,1 at the end is desired to replicate the initial phases and the gradient of these layers may not be learning well. Therefore, these gradients are likely to vanish as the network depth and the activation function are shifting their values to zero. Hence, we need to design the activation functions such that they would not move the gradient towards zero. Further, to keep the gradients in the same direction, the activation function should be symmetrical at zero. Also, the activation function should not be computationally intensive because it must be applied in each layer to calculate and takes many times in the neural networks. Most importantly, the activation functions should be differentiable because the artificial neural networks are learned through gradient descent.

In summary, the activation of a neuron is determined by the activation function. It decides the importance of input to predict the output and transforms the weighted sum of the input of the nodes in a layer into an output. Thus, during the forward propagation, the activation function adds non-linearity to the model by creating an extra task at each layer.

Generally, all hidden layers in a neural network use the same activation function. This should be differentiable to learn the parameters during backpropagation. The activation function used in hidden layers is selected considering the type of the model. In most cases, the hidden layers use ReLu activation. Sigmoid and Softmax activation is used for the output layer in binary classification and multi-class classification, respectively. The selection of an activation function should consider vanishing and exploding gradient issues.

Following are some of the widely used non-linear activation functions.

1. ReLU (Rectified Linear Unit)
    The rectified linear activation function (ReLU) is a default activation function that activates only a set of neurons at a time. It is a piecewise linear function. If the input is a positive value, the function outputs the input directly. If the input is a negative value, it outputs zero representing deactivation. As shown in Figure 2.8, this gives the $max(x, 0)$. Thus, the weights and biases for some neurons do not update during the backpropagation, which increases the efficiency of the computation process.

    Because of its characteristics such as easy computation, and not saturating, ReLU is commonly used in hidden layers. Following are the advantages (+) and disadvantages (–) associated with the ReLu function.

    + Computationally efficient, as only a set of neurons are activated.
    + The associated linear and non-saturating nature accelerates the convergence of gradient descent towards the global minimum of the loss function.
    + Since the derivative is 1 or 0, the weight is updated and reaches convergence; hence does not cause a vanishing gradient problem.

    – Dying ReLU problem (dead activation function). Since the derivative of ReLU is 0 or 1, if a derivative is zero, then the new weight is the same as the old weight.
    – Some neurons do not learn as ReLU is not symmetrical around zero and the output is zero for all negative inputs.

2. Leaky ReLU

Leaky ReLU is an extension of ReLU. It addresses the dying ReLU problem as it contains a slight positive slope for the negative region as shown in Figure 2.9. The leaky ReLu design increases the range of the ReLu activation function to become

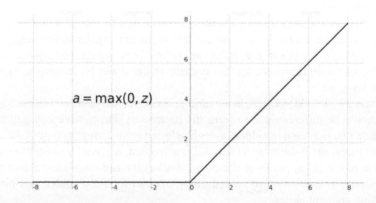

FIGURE 2.8 ReLU activation function.

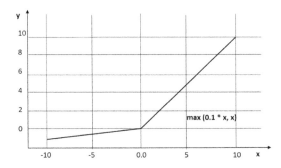

FIGURE 2.9   Leaky ReLU activation function.

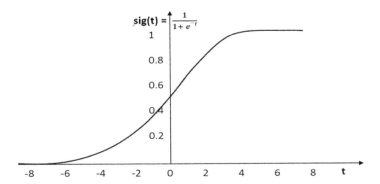

FIGURE 2.10   Sigmoid activation function.

more responsive towards the negative inputs to solve the limited range and impulsive behavior. Further, it preserves the monatomic and differentiable nature of the ReLu.

3.   Sigmoid/Logistic Activation Function

For any real value input, the sigmoid function outputs values between 0 and 1. When the input is more positive, the output becomes closer to 1. For negative inputs, the output becomes closer to 0. This function is mostly used for the output layer in binary classification. Generally, in a normal distribution, (or Gaussian distribution) the data is zero-centered with mean 0 and variance 1, which resulted in a bell-shaped curve. However, as shown in Figure 2.10, in sigmoid, the data is not 0 centered; hence consumes more computational time and takes more time to reach the convergence and to reach global minima.

Following are the advantages and disadvantages of the sigmoid function.

+ Commonly used for models that need to predict the probability as an output. Since the probability of anything exists only between the range of 0 and 1, sigmoid is the right choice because of its range.

+ The function is differentiable and provides a smooth gradient in S-shape by preventing jumps in output values.

– The sigmoid function gets negligible gradients for values higher than 3 or lower than 3. Thus, the model does not learn and faces the vanishing gradient problem, when the gradient value is closer to 0.
– Difficult to train and the model becomes unstable since the output is not symmetric around zero and the output of all the neurons will be of the same sign.
– Takes more time to converge.

4. Softmax Activation Function

The Softmax function is used in the last layer of the neural network with multiclass classification. This is described as a set of sigmoid functions, which returns the probability of each class. This activation function is more of a generalized form of the sigmoid function and produces values in the range of 0–1. It transforms the (unnormalized) output of K units of a fully connected layer to a probability distribution (a normalized output).

5. Tanh Activation Function

Tanh activation function, as shown in Figure 2.11, is a more effective function than ReLu and sigmoid, since it is zero-centered. In the tanh graph, the negative and zero inputs are mapped to strongly negative and near zero, respectively. This is mainly used in feed-forward neural networks with binary classification. The tanh function is mainly used for classification between two classes. The advantages and disadvantages of the tanh activation function are as follows.

+ Since the function output is zero-centered, the output values are mapped as strongly negative, neutral, or strongly positive.
+ The values fit between –1 and 1, tanh is generally used in hidden layers. Since the mean for the hidden layers becomes 0 or close to 0, it supports the center of the data and makes the learning process of the next layer easier.

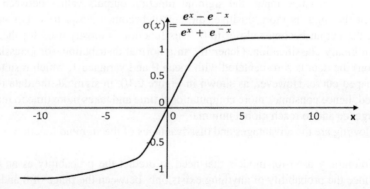

FIGURE 2.11 Tanh activation function.

+ Although this faces vanishing gradient issues since the function is zero-centered, the gradients can move in certain directions. Thus, widely used in practice.

− The gra dient of the tanh activation function faces the vanishing gradient problem. This gradient is much sharper than the gradient of the sigmoid function.

2.4.4 LEARNING RATE

The learning rate is a hyperparameter that impacts the training of a model. It addresses the model change, as a reaction to the estimated loss when the model weights are updated. Thus, the learning rate supports adjusting the weight updates, such that it reduces the loss. A traditional default value for the learning rate is 0.1 or 0.01. When the learning rate is set to very low, then the training process progresses slowly with small updates to the weights, as shown in Figure 2.12. Similarly, when the learning rate is very high, the loss function behaves differently.

Generally, the optimal learning rate depends on the model architecture and dataset. Finding the optimal learning rate supports improving performance or speeding up the training process. We need to find the optimal learning rate, in a way that minimizes the loss. For example, in each mini-batch, the learning rate can be gradually increased linearly or exponentially and calculate the loss in each increment. When the learning rate is very low, the loss value is also reduced at a slight rate. When the model enters the region of optimal learning, the loss function will show a quick drop. Accordingly, when the learning rate is increasing again, the loss value will bounce and increase again while divergent from the minimal point. It is worth noting that the slope of the curve needs to be analyzed as the optimal learning rate is associated with the sharpest drop in the loss. As shown in Figure 2.13, it is required to set the range of learning rate boundaries in a way that can observe the low, optimal, and high learning rate regions.

2.4.5 LOSS FUNCTION

The loss or error function measures the deviation of the estimated value from its true value. This indicates the status of the model's ability to predict the expected outcome. Once the loss function is defined, we can optimize the algorithm to minimize the

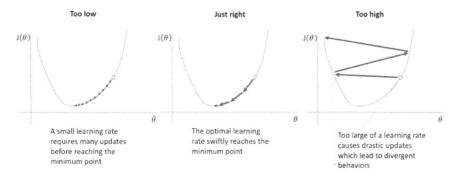

FIGURE 2.12 Types of learning rates.

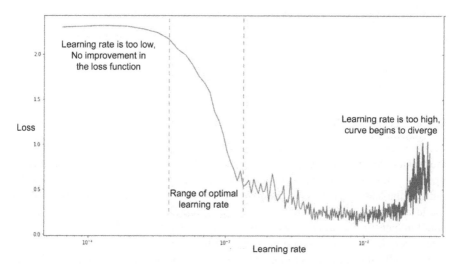

FIGURE 2.13   Setting up learning rate boundaries.

loss function. Usually, the loss is a non-negative number. Normally, better results are obtained with small loss values and the perfect predictions experienced with a loss of zero. The prediction error can be three types, namely bias error, variance error, and irreducible error that occurs due to unknown variables. Following are some of the types of loss functions associated with classification and regression problems.

For classifications problems:

1.   Binary cross entropy: use in binary classification.
2.   Categorical cross-entropy: used in multi-class classification.
3.   Sparse-categorical cross-entropy: use when the targets are integers.

For regression problems:

1.   Mean squared error (MSE).
2.   Mean absolute error (MAE).
3.   Huber loss.

During the training process, we need to find the weight vector and bias, which minimizes the total loss across all data points. Figure 2.14 shows an example of a neural network. Let y, $\hat{y}$ be the actual and predicted output, respectively. Cross-entropy is a widely used method to calculate the loss in classification applications, it reduces the distance between the predicted and actual outputs. The loss function can be calculated as in (2.4).

$$\text{Loss} = -\frac{1}{output\ size} \sum_{i=1}^{output\ size} y_i . \log \hat{y}_i + \left(1 - y_i\right).\log\left(1 - \hat{y}_i\right) \tag{2.4}$$

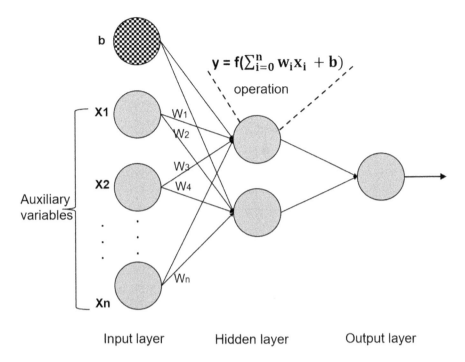

FIGURE 2.14 Example of a neural network.

When we consider regression problems, the squared error is used widely to calculate the loss. It calculates the square of the variation between the actual and predicted value. Figure 2.15 shows the graph of a regression problem for a single dimension input.

Mean squared error (MSE) is a widely used method that results in only a global minimum. That is, MSE does not get any local minima. For n number of data points, we can define MSE as in (2.5). Since it calculates the square, it prevents getting large errors. However, it does not perform robustly with outliers.

$$\text{MSE} = \frac{1}{n}\sum_{i=1}^{n}\left(Y_i - \hat{Y}_i\right)^2 \tag{2.5}$$

Mean absolute error (MAE) is a variation of a loss function in regression problems. It calculates the average of the absolute distinction between the actual and predicted values as in (2.6). Compared to MSE, MAE performs well with outliers and does not have local minima. However, it is computationally expensive.

$$\text{MAE} = \frac{\sum_{i=1}^{n}\left|Y_i - \hat{Y}_i\right|}{n} \tag{2.6}$$

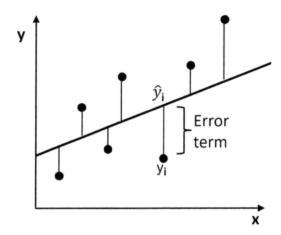

FIGURE 2.15   Regression problem representation for single dimension input.

Huber loss is another loss function that considers both MSE and MAE, which is a combination of linear and quadratic equations. As in (2.7), if the absolute value of the error a is very small, then the square of the error is divided by 2. Otherwise, it multiplies the error value with the delta. It supports better regression and works well with outliers, as the delta value solves the outlier problem.

$$L_\delta(a) = \begin{cases} \dfrac{1}{2}a^2 & for\,|a| \le \delta, \\ \delta\left(|a| - \dfrac{1}{2}\delta\right), & otherwise. \end{cases} \qquad (2.7)$$

2.4.6   OTHER HYPERPARAMETERS

The number of epochs: this defines the number of iterations of the learning algorithm during the dataset training. Thus, an epoch indicates one cycle of the training dataset, where each data sample updates the internal parameters. A training process consists of many epochs.

Dropout rate: dropout is a regularization technique used in deep neural networks to prevent overfitting. The dropout rate hyperparameter indicates the probability of ignoring a neuron during a training iteration. The dropout rate typically ranges between 0.1 and 0.5. A high dropout rate can result in underfitting, where the model does not learn well from the data. A low dropout rate can result in overfitting, where the model learns too much from the training data and performs poorly on unseen data. Therefore, the appropriate dropout rate depends on the specific dataset and architecture of the neural network.

2.5   MODEL TRAINING

2.5.1   MODEL SELECTION

Model selection identifies the best model from a set of models. Deep learning models can have different interpretations based on the various criteria that we use to define the best model. The first thing would be selecting the best hyperparameters for the model. As we discussed, hyperparameters are parameters that feed into the model learning function as input. Selecting the right hyperparameters for a model is a crucial point that affects the model's performance. Another aspect is to select the best learning algorithm for the model. Here, we need to carefully select the algorithm based on different criteria, such as the nature of the training dataset, interpretability of output, number of features, and linearity.

The most important part of the model selection comes under model evaluation. This mainly focuses on estimating the generalized error on the selected model to predict how well this model can perform on unseen data. A proper model evaluation ensures that the performance of the model will not reduce even with a completely new set of data. To do that, we need to have a completely independent test set that we have not used to train our model.

If we have plenty of available data, we can split our dataset into three main parts: training set, testing set, and validation set according to a valid ratio. The training set can be used to have different candidate models with various combinations of model hyperparameters. The models will be evaluated on a validation dataset and the best model out of all candidates will be selected. The model will be trained on a training set and validation set by tuning model parameters. The generalization error for the model is then evaluated on the test set. If this error is quite similar to the validation error, there is a higher probability that the model will perform well on unseen data as well. Even after training the model, we can further use model learning curves as a measure of model predictive performance. This will help identify overfitted and underfit models based on training and validation scores. Also, learning curves illustrate the concept of variance and bias. Bias refers to the erroneousness of the model, which will cause it to underfit the data. On the other hand, the high variance of the model will cause the model to be overfitted. Therefore, all these measurements of model complexity can be used to have a proper model selection.

2.5.2   MODEL CONVERGENCE

A model converges when additional training will not improve the model. A model converges when its loss moves towards a minimum (local or global) with a decreasing trend. Figure 2.16 shows a non-convex function with the minimum and maximum points. The term local minimum is the lowest value of a loss function in a local region. The global minimum is defined when the loss function obtains the lowest value globally across the entire domain.

In deep learning models, we need to avoid local minima. Here, the derivative with respect to the saddle point is zero, as the weights will remain the same without updating. A momentum value can be used to address the local minimum points. In

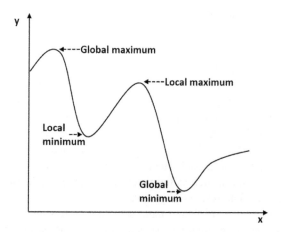

FIGURE 2.16   Optimum points.

other words, by providing an impulse in a given direction, the loss function prevents local minima points. We can use stochastic gradient descent optimization, to identify local minimum and to reach global minima, which is discussed in Chapter 5. In addition, changes in activation function, learning rate and use of batch normalization can help to avoid local minima.

2.5.3   OVERFITTING AND UNDERFITTING

In model training, we expect to have an optimal fitting, where the training error is slightly lower than the test error. The concepts of overfitting and underfitting affect low model performance. In overfitting, the model performs well on the training dataset but fails to perform on new data from the problem domain. That is, model underfitting performs poorly on both training and testing datasets. Figure 2.17 represents the underfitting, optimal and overfitting scenarios for different model types.

Before diving into the details of these concepts, we will refresh our knowledge of bias and variance. Bias is a systematic error that skews the output in favor or against the expected value. If a model gives a low error for training data, and if the model gives a high error for the test data, this can be explained as the variance. The performance metrics, such as accuracy and loss, are used to identify downsides in training, such as overfitting and underfitting.

• Model Overfitting

Overfitting occurs when the learning algorithm attempts to fit into all the data points or more than the required data points within the dataset. As a result, the model learns noise (irrelevant and unnecessary data that cause reduced model performance) and inaccurate features in the dataset that negatively affects the overall performance of the model. That means, when the model is trained with data, it picks up noise or other random fluctuations in the dataset and tries to learn from them. This will reduce the

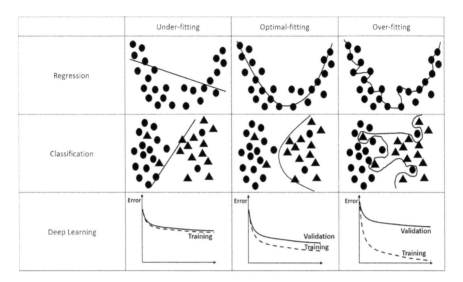

FIGURE 2.17   Underfitting, optimal-fitting, and overfitting.

model's ability to generalize and perform well with unseen data because of noise and too many details. Therefore, an overfitted model will perform well on training data but does not perform well with unseen data as shown in Figure 2.17 deep learning category. These models have low bias and high variance. Consider overfitting in a regression model in Figure 2.17. The model itself tries to cover all the data points in the graph. You may think this is very efficient, however, regression aims to find the best-fitted line and not to cover all the data points. Therefore, this model will not perform well on unseen data. Thus, an overfitting model is observed by having a low training loss, which is lower than the test loss and has a high variance.

In the early days of neural networks, the received wisdom that one cannot have more parameters than training examples. With the technology of deep learning, that rule seems to have been thrown out of the window. If there are more parameters than samples, then the model becomes overfit for training data. Then it will lose its ability to generalize data. That is, it will remember the training data very well and work for training data, but it will not give good results for a new set of related data. Since we already know the reasons for a model to be overfitted, let us learn about mechanisms that we can use to avoid overfitting in learning models.

The overfitting reduction methods are listed as follows.

- Perform regularization.
- Increase training dataset size.
- Lessen model complexity: keep the model simpler by using fewer variables and parameters. This removes noise associated with the training set to reduce variance.
- Perform early stopping in model training, when the loss starts to increase, as shown in Figure 2.18.

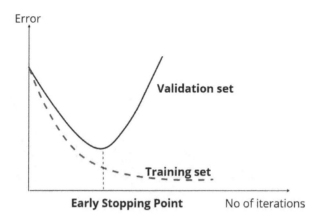

FIGURE 2.18   Early stopping in model training.

- Use dropout while training: it will drop out some of the connections and nodes randomly during training. Thus, becomes a simple and small network.
- Apply regularization, such as ridge regularization and lasso regularization, which ignore some model parameters that cause overfitting.
- Evaluate with cross-validation techniques.

When explaining the possible methods to reduce overfitting, training with more data will help to avoid model overfitting to some extent, as it provides underlying patterns of data for the model to learn. Different augmentation methods, such as cropping and rotating, can be used to increase the dataset size, which can address overfitting. However, if we add more noisy data to the dataset, this technique will not help to avoid model overfitting. Thus, it is necessary to ensure that the additional data is clean and relevant. Moreover, the number of features in the training data can be reduced to decrease the complexity of the network.

Cross-validation is an efficient method to avoid model overfitting by generating multiple train–test splits. In k-fold cross-validation, we partition the dataset into k subsets and iteratively train the model on k-1 folds while keeping the remaining fold for testing. This method helps to tune the model hyperparameters on the training data and keep the test data as an unseen set of data to select the best final model. You can read more on cross-validation in Chapter 7.

Regularization is another technique that is used to avoid overfitting in models by reducing the complexity of the model. The regularization method will depend on the type of learning algorithms that we use. As an example, we can use dropout layers on neural networks, pruning on decision trees, and penalty parameters in regression cost function and sparsity. The dropout regularization randomly ignores instances with unusual dependences from the hidden layers in model training. More details on regularization techniques will be discussed later in this chapter and Chapter 6.

As shown in Figure 2.18, early stopping can also efficiently prevent model overfitting. During model training, we measure model performance in each iteration. Therefore, up to a certain iteration, the next iteration will improve the model performance. After that point, it will weaken the model's ability to generalize on unseen data and start to overfit. This concept of early stopping refers to stopping the model training process before our learning algorithms pass through that point.

- Model Underfitting

Underfitting happens when the model is not capable of capturing the underlying trend of data and fails to learn the patterns in the dataset. In this case, the model does not learn well from the training dataset, which causes it to reduce accuracy and make unreliable predictions on test data. Usually, this occurs when there is not sufficient data to train an accurate model and attempting to train a linear model using non-linear data. In underfitting, the error or loss in both training and testing is high. The underfitted model has low variance and high bias.

Following are some of the possible methods to decrease underfitting:

- Expand the model complexity.
- Enhance the number of features.
- Use better feature extraction methods.
- Eliminate noise in the dataset.
- Increase the number of epochs to train more time.

2.5.4 REGULARIZATION

Regularization allows slight changes to the learning model for better generalization and fitting the function suitably on the training set by avoiding overfitting. This reduces the variance without considerable growth in the bias. This result in an increase in model performance on the unseen data. Although regularization improves the reliability, speed, and accuracy of convergence, it is not a solution to every problem. Figure 2.19 shows an example of underfitting, optimum and overfitting of a

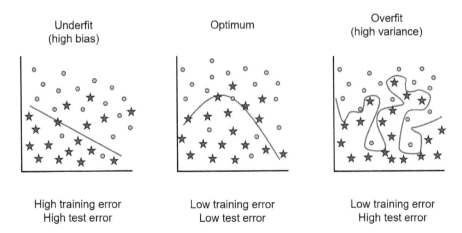

FIGURE 2.19   Example of underfitting, optimum, and overfitting of a classification task.

classification task. Generally, regularization should be used when working with large datasets in neural networks.

The widely used regularization methods are L1 regularization (lasso regression) and L2 regularization (ridge regression). L1 and L2 tend to shrink coefficients to zero, and evenly, respectively. Therefore, L1 regularization is used to select features, as the variables associated with coefficients that tend to zero can be dropped. In contrast, L2 is useful when the features are collinear or co-dependent. Additionally, L1 and L2 calculate the median and mean of the data, respectively. In a multi-layer neural network with many layers, underfitting will not happen. Due to the associated different weights and bias, overfitting can happen, as the weights are trained to fit the training data perfectly. Therefore, dropout and regularization (L1 and L2) are used as a solution for overfitting in multi-layer neural networks.

During model training, the following aspects can be observed regarding the dropout layer and regularization.

Large weights in a neural network are a sign of a more complex network that has overfit the training data.

Probabilistically dropping out nodes in the network is a simple and effective regularization method.

A large network with more training and the use of a weight constraint is suggested when using dropout.

2.5.5   NETWORK GRADIENTS

In neural network training, the associated weights are updated with small and controlled values by considering the gradients.

Generally, these models are trained using gradient-based methods and backpropagation by finding partial derivatives by navigating from the last layer to the first layer using the chain rule. Here, it is necessary to calculate the gradient from each sample element to determine a new approximation of the weight vector. Backpropagation fine-tunes the weights based on the loss value obtained in the previous epoch. It is used to calculate derivatives quickly. Thus, the weight tuning results in minimal loss and high reliability by improving the generalization. The chain rule supports finding the derivative of composite functions, or functions that are made by combining one or more functions. As stated below, it is computed extensively by the backpropagation algorithm to train feedforward neural networks. The chain rule indicates that the derivative of y with respect to x is equal to the product of the derivative of y with respect to u and the derivative of u with respect to x, as in (2.8).

$$\text{Chain rule: } \frac{dy}{dx} = \frac{dy}{du} \cdot \frac{du}{dx} \qquad (2.8)$$

However, there are issues associated with the activation functions and weight initialization mechanisms. A prominent artificial neural network design problem is the vanishing or exploding gradients and it has been there as a large barrier in model training.

(1) Exploding Gradient Problem

Consider a model with n number of hidden layers and n derivatives that multiplies together. If the derivatives are substantial because of higher weights or activation functions, then the gradient will grow exponentially as the model propagates until they explode. That is, there is a significant difference between the new weight and the old weight; thus, the model will not converge and mark different points in the gradient descent. We call this problem an exploding gradient problem.

In terms of the exploding gradient problem, the model becomes unstable and may not learn the patterns efficiently. As shown in Figure 2.13, if the learning rate is too high, it causes drastic updates without converging to a global minimum. In other words, with large changes in extreme values, the weights become large. This causes an overflow of multiplied values resulting in many weight values with missing data (NaN), which cannot be updated.

The exploding gradient problem can be detected using the following observations.

- The model does not learn much from the training set; thus, it has a low performance.
- In each weight update, the model shows substantial differences in the loss, due to the model's instability.
- In the training process, the weights increase exponentially and result in large values.
- The derivative values become constant.

(2) Vanishing Gradient Problem

In the issue of vanishing gradient, the model has slight derivatives, which cause the gradients to reduce exponentially, and the model propagates until it vanishes. Here, the accumulation of small gradients results in learning insightful patterns because the weights and biases in the initial letters are responsible for learning those core features effectively.

With the sigmoid activation function, when the number of layers increases, the derivative value that is used to update the weights becomes very low. In this problem, the derivative of the sigmoid function ranges between 0 to 0.25. Thus, the weight updating happens very slowly (due to very small derivatives) in backpropagation. Therefore, the convergence will not happen towards global minima, and sigmoid is not used for hidden layers. In the extreme case, the gradient will be 0, where the weights remain the same, and the model will stop learning.

The vanishing gradient problem can be identified using the following observations during model training.

- During the training phase, the model improves slowly and there is a possibility to stop training early. That means further training does not result in model improvement.
- The weights nearer to the output layer gets more changes and the layers closer to the input layer may not alter much.

- During model training, the weights decrease exponentially and become very small.
- During the training, the weights become zero.

Now, let us see how we can address the issues related to exploding gradient problems and vanishing gradient problems.

1. Reduce the number of hidden layers: This solution is applicable for both exploding and vanishing gradient problems. However, the model complexity reduces by lessening the number of layers.
2. Gradient clipping: This solution can be applied for exploding gradients, where the gradient size is limited to a specific range if the gradient exceeds an expected range of values.
3. Weight initialization: Applying a careful weight initialization approach would address these two issues in random initialization. Therefore, He initialization or Xavier initialization can be followed or modified if needed by adjusting the mechanism to be compliant with the data.

For further explanation, consider the sigmoid activation function and its derivative shown in Figure 2.20. Here, some activation functions squeeze the input space into an output region between 0 and 1. Thus, when the value of the sigmoid function is very high or low, the resulting derivative output becomes low. This result in vanishing gradients and low model performance. However, as the number of layers is increased, the gradients become very small and perform effectively.

Usually, when the model has a smaller number of layers, sigmoid activation function is used. Therefore, when there are many layers, a sigmoid is not used for each layer. In such scenarios, activation functions such as ReLU, which does not result in a small derivative, are used. As another solution, residual models can be used, as they provide residual connections straight to earlier layers. The most recommended approaches to overcome the vanishing gradient problem are layer-wise pretraining.

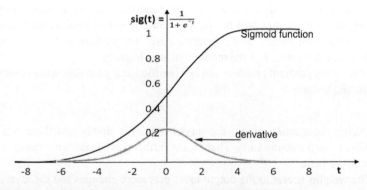

FIGURE 2.20   Representation of sigmoid function and its derivative.

REVIEW QUESTIONS

1. What is the use of loss function in neural networks?
2. How can hyperparameters be trained in neural networks?
3. Why do we use activation functions in deep learning models?
4. How do you initialize weight and biases in neural networks?
5. What are the limitations in zero initialization of weight?
6. Why is ReLU the most commonly used activation function?
7. Justify the cases where the linear regression algorithm is suitable for a given dataset.
8. List some of the metrics used to evaluate a regression model.
9. How can gradient exploding and vanishing problems be resolved?
10. Show the relationship between the model complexity with the training and test error in overfitting and underfitting scenarios.
11. How can if the model is overfitted or underfitted using learning curves be identified?

3 State-of-the-Art Deep Learning Models: Part I

3.1 OVERVIEW OF NEURAL NETWORKS

This chapter discusses different deep learning techniques and their design concepts. First, we need to understand the need for deep learning algorithms over machine learning. For that, it is important to identify the advantages of neural networks compared to traditional machine learning techniques, since neural networks require more computational power. Let us see the interaction of different types of neural networks with real-world applications. In Chapter 2, we discussed the composition of neural networks that consists of interconnected layers of nodes in a way that mimics human neurons. We learnt that neural networks perform mathematical operations to detect patterns in data. The main reasons for using deep learning over machine learning can be listed as follows.

- Decision boundary: a classification algorithm learns the model to determine the class, which a given data point belongs. For example, consider the logistic regression shown in Figure 3.1. Here, the sigmoid function is used to separate the data points into two classes with a linear decision boundary. It does not support learning decision boundaries for non-linear data. Accordingly, machine learning algorithms cannot learn all the functions. Therefore, not all problems with complex relationships can be solved by machine learning algorithms. Hence, deep learning techniques have emerged.
- Feature engineering: this process consists of feature extraction and feature selection. For instance, consider an image classification problem. The manual feature extraction process requires strong knowledge of the subject and the domain of the image and consumes more time and effort. However, deep learning techniques automate the feature engineering process as shown in Figure 3.2.

The main types of deep learning models are as follows:

- Artificial neural network (ANN): use to solve regression and classification tasks.
- Convolutional neural network (CNN): use for the classification of image and video data. Mainly for object recognition, object detection, and object

DOI: 10.1201/9781003390824-3

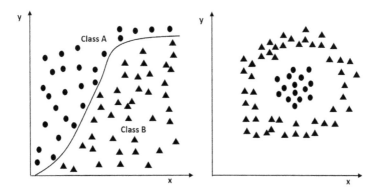

FIGURE 3.1 Decision boundary of linear (left) vs non-linear data (right).

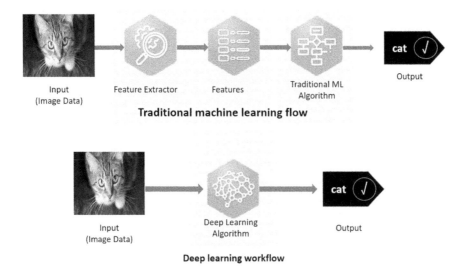

FIGURE 3.2 Comparison of machine learning and deep learning flows.

classification. CNN has additional layers for convolution and max pooling, compared to ANN.

- Recurrent neural network (RNN): data flows in any direction and is used for applications, such as object detection, and language modeling. Long short-term memory is effective for this use, which involves embedding layers and one-hot implementation.

3.2 ARTIFICIAL NEURAL NETWORKS

Artificial neural networks (ANNs) are designed to simulate the human brain. It consists of input space, hidden layers, and output layers with connected nodes to simulate the human brain. In a node, the weighted inputs are summed up, and an

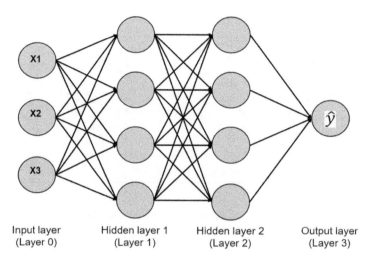

FIGURE 3.3 Example of an ANN.

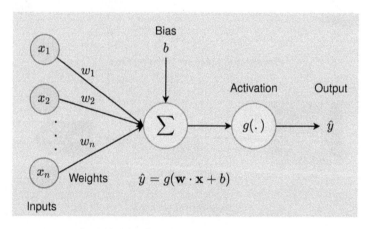

FIGURE 3.4 Operation of a perceptron.

activation function is applied to get the results as shown in Figure 3.3. The input layer gets the inputs from the input space and passes to the hidden layer to process where each layer learns certain weights, and the output layer delivers the result. This is a feed-forward network, where the data moves through the input, hidden, and output nodes without any loops in the network. These are mainly used for regression and classification of tabular data, image data, and text data.

In an ANN, a node or perceptron performs a mathematical operation as in (3.1), where x is the input, w denotes the associated weight, and b is the bias. As shown in Figure 3.4, the input is multiplied with the corresponding weights, and the summation and the bias is added to the result. Finally, an activation function g is applied resulting in an output of $g(\mathbf{w} \cdot \mathbf{x} + b)$.

$$x \mapsto \sum_{i=0}^{n} w_i x_i + b = w_0 x_0 + w_1 x_1 + \ldots + w_n x_n + b = w.x + b \qquad (3.1)$$

There are many advantages associated with ANNs. ANN supports learning any non-linear function and considers it as a universal function approximator. The non-linear characteristics of the model are handled by the associated activation functions, which learn the complex relationship. Here, each node outputs a weighted sum of inputs. If a model does not associate with an activation function, then the model can learn only the linear relationships. Thus, the activation function gives power to ANN.

Let us consider the challenges in ANNs. Consider an image classification problem. Initially, the two-dimensional image is transformed into a one-dimensional vector before the training starts. However, when the image size increases, the required number of trainable parameters also increases. For example, consider an image of size 224×224. In this case, the first hidden layer with four nodes will contain 602112 trainable parameters. Therefore, ANN drops the spatial features and the pixel arrangement of the image. Further, ANN may not capture the sequential data in the input space. This limitation is addressed using recurrent neural networks (RNNs).

3.3   RECURRENT NEURAL NETWORK (RNN)

Recurrent neural networks (RNNs) are derived from feed-forward networks and used to learn sequence data. RNN is a type of ANN, where the nodes are connected maintaining a temporal sequence. As shown in Figure 3.5, RNN has a looping connection in the hidden layers to capture the sequential information in the input. Here, the output of a given layer is fed back to the input as feedback, and then the output is predicted as shown in Figure 3.6. Generally, RNNs are used to solve problems related to time series data including both text and audio. RNN-based models are mainly used in speech recognition, natural language processing, language translation, stock forecasting, and image captioning.

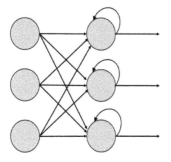

Recurrent Neural Network

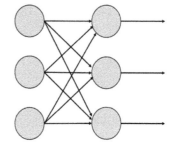

Feed Forward Neural Network

FIGURE 3.5   RNN vs ANN.

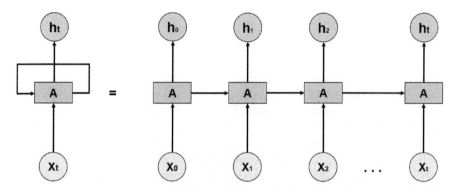

FIGURE 3.6   Recurrent neural network.

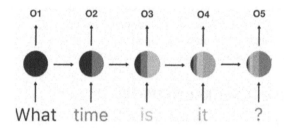

FIGURE 3.7   Sequence data processing in RNN.

Generally, the deep feed-forward models may need specific parameters in each element to handle sequence data and are not capable of generalizing to variable-length sequences. On the other hand, RNNs efficiently work with sequential data by memorizing part of the input data and using them to make accurate predictions. The advantage is that RNN attains the sequential information in the input space. Thus, as depicted using different color codes in Figure 3.7, it considers the dependency among the words in the prediction process. In this example, the outputs o1, o2, o3, and o4 at each time step are based on both the previous and current words.

Generally, in sequential data processing, it is necessary to consider the order of elements in the sequential data. As an example, when considering sequential data, such as natural language processing, it has input data in a sequence form and maybe the output data also in the sequence of elements. In this case, the order of sequence elements is important to have good predictive results. If we are using MLP to process this data, one solution would be to have input layers that contain a set of units equal to the set of elements in the input sequence. However, this will not work with variable-length sequences. Additionally, these input elements need to be in the same order. Sometimes, the same sentence may be expressed in various ways with different word reordering. Therefore, without a dataset that contains all these orderings for similar meanings, it will not be effective to learn an accurate model.

Recurrent neural networks are mainly designed to overcome problems in dealing with sequential data. In RNN, we have element-level models that are finally used together to form the final predictive model. The speciality of these models is that all element-level models share the same parameters. Therefore, it does not matter where the element is present in the sequence, as all of them are processed in the same manner and remain in the order in the outputs. Thus, each of these element-level models takes input from each element and the output of the element-level model of the previous element in the input sequence. After progressing in this way and after processing the final element in the sequence, the sequence data can be encoded to output the final element-level model. These output data can also be decoded into sequential outputs, such as speech recognition, and language translation.

Consider the property of parameter sharing across different time steps. The recurrence relationship to obtain a full chain of input is known as unfolding the equation. This shows the hidden state at any given time t as a function of parameters and sequence. With this, we can use fewer parameters for training, hence reducing the computational cost. As shown in Figure 3.8, U, W, and V are the three weight matrices that are shared throughout the time steps. Thus, the same weight propagates forward direction over time and the weights get an update during backward propagation.

In order to train an RNN model, we need to infer the parameters. That means, during the forward pass, the gradients of the models need to be derived by calculating through all the hidden states. This is known as unrolling the computational graph to compute hidden states and then using it to compute the gradients. These gradients are calculated using backpropagation, by sequentially going backwards in time starting from the gradient of hidden variable $h^{(t)}$ to $h^{(1)}$, which is known as backpropagation through time.

When we talk about the challenges, vanishing and exploding gradient problems occur in deep RNNs with many time steps. During the model training, the error is calculated using the cost function at each time step and used backpropagation to update the weights. Therefore, every single neuron is associated with updating its weights to minimize the error. This vanishing gradient problem occurs when the error is moving backwards through all the neurons to get their weights updated. In RNN, the cost function that is used in a given time state will be used by other shallow layers to update their weights. Thus, the gradient value that is calculated at each step will be multiplied back through the weights earlier in the network. This causes the gradient to vanish if the weight is too small as the gradient becomes less and less with each

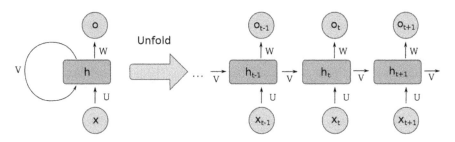

FIGURE 3.8 RNN architecture.

multiplication. Based on the value of the weight *W*, this causes two main problems as follows:

- When *W* is too small, this will cause a vanishing gradient problem.
- When *W* is too large, this will cause an exploding gradient problem.

As solutions to vanishing gradient problems, we can initialize weight to the neural network such that to prevent backpropagation to have unrealistically small values for weights. Additionally, we can use echo state networks, which is a specific neural network that is designed to avoid vanishing gradient problems. Long short-term memory networks (LSTMs) are also a better solution to this problem. In order to avoid exploding gradient problems in RNN, we can stop the backpropagation at a certain time step before the exploding gradient occurs. Penalizing will also be a solution to this problem to reduce the impact of backpropagation with la. Another solution would be to put a maximum limit on gradients using gradient clipping.

3.4   CONVOLUTIONAL NEURAL NETWORKS

3.4.1   OVERVIEW OF CONVOLUTIONAL NEURAL NETWORK

Convolutional neural networks (CNNs) are widely used across different application domains and are mainly applied for image and video classification. CNNs automatically detect the essential features. For example, among many pictures of flowers and trees, CNN learns unique features for each class. The architecture of CNN is similar to the connectivity pattern of biological neurons. A given neuron is responsible to identify a particular area of the image and a set of such regions overlaps to cover the entire visual area. Generally, CNNs require less preprocessing and are computationally efficient as they are not densely connected. Here, only a set of input nodes impact the corresponding outputs, enabling flexible learning. Also, having a small set of weights in a layer allows the processing of high-dimensional inputs, such as images.

Kernels, also known as filters, are the main building blocks of CNNs that use convolution operations to extract the appropriate features from the input and produce a feature map. A CNN consists of a stack of layers associated with different kernels that extract spatial patterns such as edges by detecting the intensity changes in the image. It can capture both spatial and temporal dependencies in the image using the relevant filters. With the concept of parameter sharing, a given filter is applied over the regions of the input to generate the feature map. Figure 3.9 shows the importance of filters in the feature extraction process of images. The low-level features are considered as general features, such as lines, corners, and edges. The mid-level features can be named as object parts with texture and structure. The high-level features are the whole object to identify details of the category. Thus, CNNs perform well in computer vision and image classification.

As discussed, there are many advantages associated with CNNs. CNN captures the spatial features that correspond to the image pixel arrangement and associated relationships. It allows accurate object recognition, its location, and its relationship

low-level features   mid-level features   high-level features

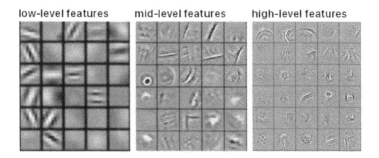

FIGURE 3.9   Output of convolution.

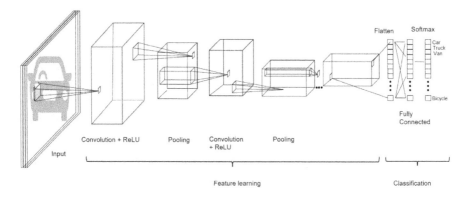

FIGURE 3.10   Layers of a CNN.

to other objects in the image. Accordingly, CNNs solve problems related to both images and sequential data. Although CNN works for small size images, it does not ensure high precision. Because, when the image data is flattened into an array, it loses important dimensionality information or spatial information of the image. CNN can identify image complexities by reducing the number of parameters and reusing the weights. Thus, CNNs provide high-performance results. As shown in Figure 3.10, a CNN consists of four main layers namely the convolutional layer, pooling layer, ReLU correction layer and the fully connected layer. The details of each layer are described later in this chapter.

3.4.2   CONCEPTS OF CNN

Let us discuss some of the concepts and terminology used in CNNs.

- **Batch normalization** increases the efficiency of a neural network. It also increases the stability of the model by normalizing the inputs for the layers using recentering and rescaling. This stabilization of the learning process reduces the required epochs for the training.

- **Batch-norm layers** enable independent learning of each layer of the model. It normalizes the output of the former layers supporting efficient learning. This is used as a regularization to prevent overfitting.

- **Dropout layer** activates and deactivates hidden layers by randomly setting inputs to zero, to prevent overfitting. Here, the applicable layers are set to True, so that the associated values will not drop during training. The outputs of a layer under dropout are randomly sub-sampled, thus having the effect of reducing the capacity. That is, it randomly selects only a set of features by activating only some activations with the input. Normally, weights are updated by multiplying with the dropout ratio, which is between 0 and 1. With proper hyperparameter optimization, we normally select a value higher than 0.5. Usually, dropout is placed on the fully connected layers only because they contain more parameters. However, since it is a stochastic regularization technique, this can place everywhere.

- **Attention layer** lays the corresponding elements that interact with each other when combining multiple vectors. The attention layer enables meaningful fusion by laying elements differently.

- **The dense layer** has deep connections such that each node gets inputs from all the nodes of its former layer. It uses a linear operation resulting in each output being generated by a function based on each input. Accordingly, this layer generates features to learn from all the combinations of the features of the prior layer and widely used layer. The associated matrix-vector multiplication uses the trainable parameters and updates during the backpropagation. Usually, this layer is used to change the vector dimensions, and perform operations such as scaling, rotation, and translation on the vector. This process adds a fully connected layer to the model.

- **Filters** learn to detect abstract concepts in more depth. It identifies spatial patterns, such as edges by detecting intensity changes in the image.

- **Flatten** converts the pooled feature map to a single column and forward to a fully connected layer. It is necessary to include a flatten operation after a set of 2D convolutions or pooling, because, flattening converts the multi-dimensional data into a one-dimensional array and input to the next layer. We flatten the output of the convolutional layers to create a single long feature vector. It is connected to the final classification model, which is called a fully connected layer. A flatten layer collapses the spatial dimensions of the input into the channel dimension, as shown in Figure 3.11. The intuition would be to feed the image data, which is given as a matrix of pixel values to a flattened array of values and apply it for learning. In practice, we add two hidden fully connected layers, such that, after the convolutional layers and before the output layer. The reason is that convolutional layers try to extract features in a differentiable manner, and fully connected layers try to classify the features. Using two dense layers is more advised than one layer.

- **Tensor** is a multidimensional data structure. It is a generalization of vectors and matrices, where vectors are one-dimensional and matrices are two-dimensional data structures. For instance, second-rank tensors as matrices. Tensors have properties that not all matrices will have. In practice, we can use NumPy for working with an array and TensorFlow for working with a tensor.

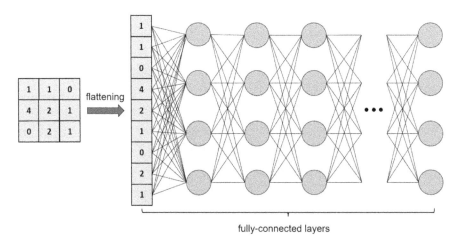

FIGURE 3.11 Flattened data representation.

- **Padding** indicates the number of pixels added to an image during filter processing. The model design becomes easier if we set the width and height of the image matrix. Thus, we do not have to be concerned about tensor dimensions. Padding allows us to design deeper networks and avoids image size decrease during the convolutional operation. This is used to get the output image size the same as the input image size, without losing any information after the convolution operation. Different types of padding are used to handle the border pixels of the image matrix. In zero padding, we add zeros to the borders systematically. Other padding types are reflection padding or mirror padding, near value padding, which is based on neighboring pixels.
- **Stride** is a kernel parameter that changes the amount of movement across the image region to reduce the image size. For instance, when the stride is set to 1, the filter moves one unit at a time. Stride controls how the filter convolves around the input volume. Figure 3.12 shows the 3×3 feature map obtained from a 7×7 input volume with stride 1 and 3×3 filters. Since neighboring pixels in the lowest layers are strongly correlated, the size of the output is reduced using sub-sampling, also known as pooling, the filter response. Accordingly, a large stride in the pooling layer results in more information loss.

Now let us dive into the components of a CNN as shown in Figure 3.13. They can be listed as

1. Convolution layer/kernel.
2. Max-pooling layer.
3. Fully connected layer.

In practice, we need to determine the number of layers, when designing a model for optimal performance. In general, the number of hidden nodes should lie in the

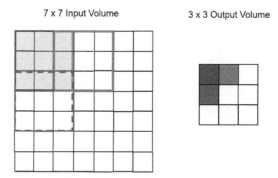

FIGURE 3.12   Stride representation.

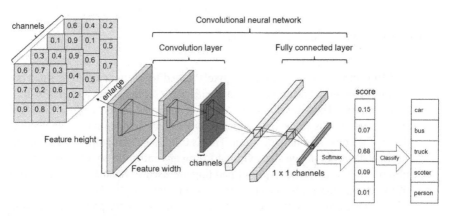

FIGURE 3.13   Example of a CNN.

range of the size of the input and output layers. One possible technique used by practitioners is the number of hidden nodes should be 2/3 the size of the total size of input and output layers. Another possible approach is the number of hidden nodes should be fewer than double the input layer size. However, based on the problem we need to experiment and identify the optimal number of hidden layer nodes for the minimum loss.

3.4.3 CONVOLUTIONAL LAYER

Convolution is a mathematical operation that interconnects the image matrix and a filter, using a function change and producing an output. It is the first layer that extracts features from the original image. Convolution learns the image features and maintains the relationship among the pixels. A convolutional layer is the main element of a CNN. This consists of filters (kernels), whose discrete value parameters (kernel weights) need to be learned. Usually, the convolutional layer applies a filter, which is smaller than the actual image or feature map size, where each filter convolves with the image and produces an activation map. At the start of the CNN training process, random numbers are allocated to the kernel weights. Numerous approaches are employed to initialize the kernel weights, which are changed on a per-training-step basis.

First, let us understand the term 'kernel', which is an operation to extract features to solve non-linear problems using a linear classifier. This function is applied to each data unit to transform the non-linear features into a separable higher-dimensional region. In detail, the matrix of the kernel moves across the input space performing the dot product with a sub-region of the input and producing the output as the matrix of dot products. More precisely, a kernel denotes a 2D array of weights and a filter represents a 3D stack of multiple kernels. Here, a kernel is allocated to a given input channel. Thus, a 2D filter is the same as a kernel. However, a 3D filter is a collection of kernels.

Consider a case where the input of the CNN is a multi-channeled image. Each CNN kernel slides across the two-dimensional input space, conducting element-wise multiplication and a summation to produce a single output in reduced size. This process is repeated until no sliding is possible and generates a two-dimensional feature map (activation map). Figure 3.14 graphically illustrates the convolution operation of a 2×2 kernel applied to a 3×3 two-dimensional input.

As we learned, a convolution presents the overlap size of a function as it blends over another function. Generally, starting from the top left corner of the image, the convolution moves the filter over all the positions, such that the filter fits with the image boundaries. As shown in Figure 3.14, the first entry of the output activation map generates by convolving the filter with the selected region of the image. This process is repeated for each element of the image to obtain the activation map. Thus, the convolutional layer's output is produced by stacking each filter's activation map along the depth, where each element of the activation map is considered as an output of a node with parameter sharing. Therefore, each node in the convolutional layer is associated with a region in the input image, and the filter size is the same as the area size. The associated local connectivity enables learning filters with higher response to a region of the input image. Generally, the basic features, such as the edges, are captured using the initial convolutional layers and the complex features, such as shapes and objects, are identified using the final layers.

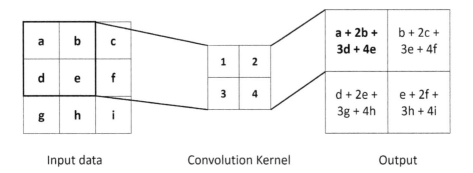

Input data   Convolution Kernel   Output

FIGURE 3.14   Convolution operation.

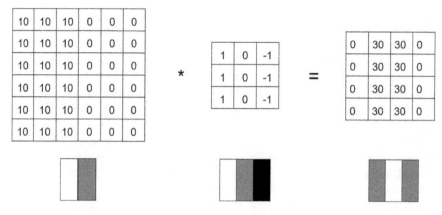

FIGURE 3.15 Vertical edge detection.

Accordingly, the convolution layer extracts features from raw data and ensures spatial connectivity among the pixels by learning image features using different regions of the image. This layer reduces the input image size while identifying the major information in the image. Filters identify spatial patterns, such as edges, using the intensity values, as shown in Figure 3.15. The kernel size is the multiplication of the width and height of the filter mask. This is done by the convolution operation, where a fixed kernel matrix is passed through the input image matrix and calculates the resulting output of the matrix. Generally, the number of weights in a convolutional layer is less than the number of weights in a fully connected or dense layer. Therefore, it is followed by a non-linear activation function. Accordingly, the convolutional layer identifies a local association of features from the prior layer and maps their presence to a feature map.

3.4.4 POOLING LAYER

The pooling layer is used to decrease the spatial size of the convolved feature by sub-sampling. The reduction of the feature map dimensions results in lowering the learnable parameters and the computational power in the model. Additionally, it extracts the main features from the feature map supporting effective training. Generally, the pooling layer is used in between two convolutional layers, as the pooling operation reduces the spatial volume of the input image after the convolution. If we use a fully connected layer after the convolutional layer without applying pooling or max pooling, it results in expensive computations. Thus, we can use a technique, such as max-pooling to lessen the image's spatial volume. Several types of pooling techniques are available, such as max pooling, min pooling, and mean pooling.

The max-pooling technique outputs the highest values from image regions covered by the convolutional kernel. The mean pooling returns the average value of all the values from the convoluted feature map. In an image, if a region impacts the presence of a given feature, the max pooling identifies that feature. Comparatively, if there are regions with contradictory presence, then mean pooling is used. From another point

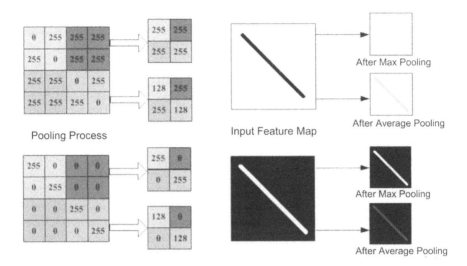

FIGURE 3.16   Max pooling and average pooling comparison.

of view, max pooling discards the noise data associated with the input image by reducing dimensionality. In contrast, average pooling performs dimension reduction as a noise-suppressing function. Therefore, the max-pooling function performs well than the average-pooling function.

Figure 3.16 shows two examples of max pooling and average pooling with 2×2 filters and stride 2. Max pooling performs well for images with a black background and white objects. Min pooling works well when the image has white backgrounds and black objects. Since average pooling smooths out the image, it is hard to identify strong features with it.

In CNNs, convolutional and max-pooling functions are going in hand in hand. The number of these two layers needed is decided considering the image complexity. Thus, the increased number of such layers is capable of capturing low-level highly complex details but we have to bear the increased computational complexity and power.

3.4.5 FULLY CONNECTED LAYER

A fully connected layer has a dense relationship such that each node is linked with each node in the prior layer and weight is associated with each connection. Generally, these layers are included before the output layer of the CNN, as they flatten the results before classification. Usually, this supports learning non-linear combinations of complex features as given by the convolutional layer.

When the CNN is learned by capturing low-level features of the input, the result can be flattened and fed to a feed-forward network. Then the results are backpropagating to continue with the training. After having multiple iterations, the model leans the dominating low-level feature associations with the labels and classifies them using an activation function such as Softmax. In summary, after learning high-level features from the convolutional and max-pooling layers, a fully connected layer is used to learn the linear combinations of the features we extracted.

3.5   COMPARISON OF ANN, RNN, AND CNN

This section states the advantages, disadvantages, and features' comparison of the discussed main three models: ANN, RNN, and CNN, as given in Table 3.1 and Table 3.2. We learnt that an ANN is the basic variant of neural networks that contains a set of neurons in each layer. This is a feed-forward network as the inputs are processed in the forward direction through the nodes in different layers.

As we discussed in this chapter, RNNs are more complex and pass the information both forward and backwards. Here, each node behaves as a memory cell and the result of a node is fed back to the model. The model self-learns and progresses using backpropagation until it predicts the correct result. When we consider CNNs, they consist of convolutional layers that are connected or pooled. These layers produce feature maps for each considered region of an image. Different types of CNN are available in practice and we discuss them in Chapter 4.

After discussing the features of each of the main neural networks, it is important to compare and contrast these models for a better understanding of their practical use. Mainly, in an ANN each node is connected to every other node. In contrast, in a CNN, only the last layer is fully connected.

In CNN we apply different filters also known as kernels. Thus, while applying kernels, it considers the neighboring pixels, which helps the network to learn

TABLE 3.1
Summary of ANN, RNN, and CNN

Model	Advantages	Disadvantages
ANN	• Store details about the entire model. • Train with partial knowledge. • Capable of error handling. • Use a distributed memory.	• Depend on the hardware specification. • Limited model explainability. • Need a well-defined model structure.
RNN	• Record each detail along the time. • Use convolutional layers to extract features. • Predicts time series data. • Suited to analyzing temporal, sequential data, such as text or videos	• Has vanishing gradient and exploding gradient issues. • Complex training process. • Hard to process long sequences with Tanh or ReLU activations.
CNN	• Perform well on image classification. • Automated feature extraction. • Parameter sharing.	• Does not consider the location and direction of the object. • Require more training data.

TABLE 3.2
Comparison of ANN, RNN, and CNN

	ANN	RNN	CNN
Type of Data	Tabular or Text Data	Sequence Data	Image Data
Spatial relationship	No	No	Yes
Recurrent connections	No	Yes	No
Static input length	Yes	No	Yes
Vanishing and exploding gradients	Yes	Yes	Yes
Parameter sharing	No	Yes	Yes
Performance	Less effective than CNN, and RNN.	RNN has less feature compatibility than CNN.	More effective than ANN, and RNN.
Application	Image classification and Computer vision.	Text-to-speech conversions.	Image classification, text digitization and NLP.
Features	Concrete data points are vital to extract features. Do not identify image features like angles, light, darkness, inversion.	Perform well with sequence data. However, less feature-compatible.	Different filters are used to detect image features. Extract the exact image features leaving the noise and improving clarity.
Scalability	Do not scale well. Trainable parameters increase drastically when the image size increases.	Do not scale well for lengthy sequences of data.	Support scalability as a result of capturing spatial features.
Computational power	Require large computation power while taking images as inputs and training.	Slow and complex training procedures.	Reduces the number of parameters: weights and bias. Weight sharing reduces the number of parameters.
Main advantages	Fault tolerance, Traub with partial details.	Remembers each detail, predict time series data.	Accurate image recognition. Weight sharing.
Disadvantages	Hardware dependence, unexplained model behavior.	Gradient vanishing, exploding gradient.	Require more training data, does not consider the location and direction of the object.

different features, eventually, these are converted into a 1D array in the feed-forward network and get the classification results. On the other hand, ANN requires fixed data points. For example, consider a model that distinguishes cars and buses. Here, the details, such as the height of the vehicle and front shape are given as explicit data points. However, a CNN extracts these spatial features from the original image. Likewise, CNN can extract many features automatically, without measuring each of the features.

Therefore, considering better feature extraction ability, CNN has different filters to detect many features of the image and have good clarity of the image as output with max pooling. It learns features from the image's region of interest (ROI). However, ANN creates the image in a multi-dimensional array and the features of the image are not recognized. This is the reason CNN is useful and extracts the exact image leaving noise. When we increase the image size, the number of trainable parameters increases drastically in ANN, failing to capture the features. However, CNN extracts the spatial features from an image. Hence, ANN does not scale well with input size and requires large computation power while taking images as inputs and training.

Further, considering RNN models, they include less feature compatibility when compared to CNN. RNN fed the same weights and bias to all layers to transform the independent activations into dependent activations. This helps to lower the complexity due to parameter increase and remembering the output from each of the prior layers that fed into the next hidden layer.

REVIEW QUESTIONS

1. Compare and contrast the following architectures:

   artificial neural networks, recurrent neural networks, and convolutional neural networks

2. Describe the main layers in a CNN and their process.
3. What is the importance of padding in a CNN?

4 State-of-the-Art Deep Learning Models: Part II

4.1 FEED-FORWARD NEURAL NETWORK

Feed-forward neural networks are quintessential deep learning models that form the basis for many related models, such as CNNs and RNNs. The main goal of a feed-forward network is to assess a given function f^* such that for a given classifier $y = f^*(x)$, where x and y denote the input and output, respectively. The mapping $y = f(x; \theta)$ is defined by learning the parameter θ, such that it results in an optimal solution. This network is known as feed-forward because data flows only in the forward direction. Therefore, the middle nodes that process the function f produce the output y without any feedback connections to the same node. Figure 4.1 shows an architecture of a feed-forward neural network with a directed acyclic graph that has input, hidden, and output layers.

These neural networks put together different functions and build a directed acyclic graph. As an example, consider the functions $f^{(1)}$, $f^{(2)}$, and $f^{(3)}$ denote the first, second, and third layer, respectively, that are lined in a chain to compose the function $f(x)= f^{(3)}(f^{(2)}(f^{(1)}(x)))$, which is a common structure in a feed-forward neural networks. The output corresponds to the final layer and the chain length is considered as the model's depth. The training of the model derives $f^*(x)$ from $f(x)$, where each value x is associated with a label y, that corresponds to $f^*(x)$.

The behavior of the other layers, except the output layer, is mainly based on the learning algorithm and not directly by the dataset. Since the training data is not directly used to get the output of these layers, they are known as hidden layers. In the general scenario, there can be multiple hidden layers, where each node in the layers represents neurons. The hidden layers are vector-valued and used to determine the depth of the model. Each hidden layer in a network represents a single vector to scalar function, which consists of many units that act in parallel. These units resemble neurons that get inputs from other units and compute their activation function. This concept of using multi-layer vector representation is derived from neuroscience with the computation of biological neurons.

A common way to understand feed-forward networks is to consider the limitations of linear models. These models are used with linearly separable data to obtain a linear boundary. However, the capacity of such models is only limited to linear functions and is not capable of understanding the intersection between input values. Therefore,

DOI: 10.1201/9781003390824-4

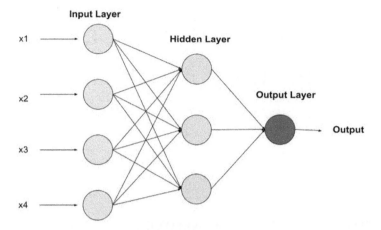

FIGURE 4.1   Feedforward network architecture.

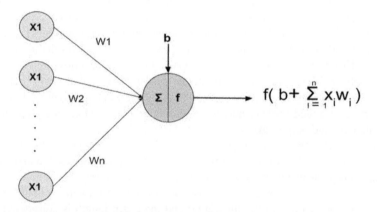

FIGURE 4.2   Neuron cell output calculation.

these multi-layered networks of neurons can handle non-linearly separable data. The complex non-linearity relations between input and output data are handled by the hidden layers.

The neurons calculate a weighted sum of inputs and then apply activation functions for the normalization of the summation. Thus, each neuron has weights that are associated with that neuron and these weights are learned during the process of training. Figure 4.2 shows the calculations associated with a neuron. The neurons used the activation function to learn the linear or non-linear decision boundaries. This also has the effect of normalizing to prevent the output of each neuron from becoming very large after several layers due to cascading effect. The widely used activation functions are sigmoid, tanh, and ReLU.

Let us consider the model learning process. When the training data are passed through the network, the actual and predicted outputs are compared. The identified value of error is used to change the weights of neurons in such a way that minimizes the error gradually. This process is done using backpropagation. Then batches of data will iteratively pass through the network by updating the weights until the error is minimized. The hidden layers in the network use a variety of functions for data transformation. Therefore, each hidden layer is focused on a given out based on a feature. Finally, the output layer gives out the prediction.

4.2 MULTI-LAYER PERCEPTRONS

A multi-layer perceptron (MLP) is a fully connected feed-forward neural network. Artificial neural networks consist of neurons, which are also known as perceptrons. A perceptron is a single-layer neural network consisting of input values, weights, and bras, a total sum, and an activation function. Perceptrons are useful in classifying linearly separable datasets and encounter problems when non-linear functions, such as the XOR function, are performed. The MLPs have the same input and output layers but may have multiple hidden layers in between the layers, as depicted in Figure 4.3. The MPLs break these restrictions and classify data that is not linearly separable; thus, supporting the binary classification with supervised learning of complex datasets.

The MLP algorithm passes inputs forwards via the network by calculating the dot product of the inputs and weights. The dot product provides a value at the hidden layer and utilizes activation functions in each layer. Then MLP passes the calculated value to the next layer and the above steps are repeated until the expected output is generated. The output will be used for either backpropagation to train the model again or for the testing process to make decisions. MLPs provide the foundation for neural network processing to enable computer vision systems to solve far more complex

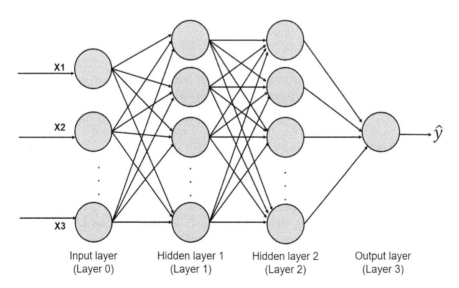

FIGURE 4.3 MLP architecture.

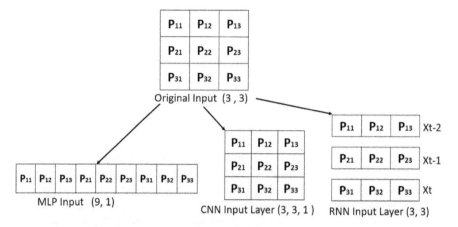

FIGURE 4.4   Layout comparison of MLP, CNN, and RNN.

problems than the XOR problem. Although MPLs are widely used for regression, they are not well-suited to identify patterns in sequential and multi-dimensional data, as the input's spatial information is not considered in MLP architecture.

Figure 4.4 shows a layout comparison of MLP, CNN, and RNN. In comparison to the CNN architecture with convolutional and pooling layers, one of the key features of MPLs is the ability to regularize the data to prevent overfitting or underfitting. The new function named dropout comes into play to resolve this problem. This layer discards a portion of units based on a random rate. Consider an example with 256 units in the first layer. If the dropout is 0.4, then only (1–0.4) * 255 = 153 units will be passed to the next layer. This technique performs well for unseen input as the model is trained in such a way that it will make correct predictions even with missing data.

MLP is now deemed not sufficient for complex computer vision applications, as the parameters grow by multiplying the number of perceptrons in each layer and creating redundancy with high dimensions. Spatial information discard is another limitation as it inputs flattened vectors. However, a lightweight MLP with two to three layers provides good accuracy levels.

4.3   GENERATIVE ADVERSARIAL NETWORK (GAN)

Generative adversarial network (GAN) is one of the recent machine learning advancements that act as generative models, where two neural networks namely generator and discriminator, contest to provide accurate predictions. GAN creates new data instances by replicating the training data. It combines a generator with a discriminator, where the generator is trained to predict the output and the discriminator is trained to differentiate the actual and predicted output. The generative models provide new data instances, whereas the discriminative models distinguish different data instances. Thus, generative models deceive a set of statistical models that contrast with discriminative models.

The generative models capture the joint probability of $p(X, Y)$, which describes the probability of having the given input vector and the label together. In contrast, the discriminative models measure the conditional probability of $p(Y|X)$, where the probability of having label Y when the input vector X is given. As we can dive more into the dataset, we can see that a generative model includes the dataset as a whole and provides how likely the given example presents. On the other hand, the discriminative models ignore the whole-dataset aspect and show the likeness of the appropriate label that is given to the data instance.

GANs are useful as an image augmentation method since they can generate a proper dataset that matches the original datasets. We learnt that the generative model produces obvious fake data instances during the training process and the discriminator quickly classifies them as fake. The generator learns to generate data instances that are getting closer to the input data so that it can mislead the discriminator. When the generator completes its learning, the discriminator fails to distinguish the actual data and the data instance generated by the generative model, hence decreasing the accuracy. The process view of the generative adversarial model is depicted in Figure 4.5. Here, the output of the generator is fed to the discriminator. The output of the discriminator, which is the classified signal is used for backpropagation to update the weights of the generator. It considers the impact of the generator weight, which depends on the discriminator weights it feeds into. Subsequently, the generator produces fake data that are similar to the input data based on the feedback received from the discriminator. The loss function of the GAN is a combination of the loss of the generator and the loss of the discriminator. Here, the discriminative loss indicates the misclassification of real data as fake data.

However, the training of the generator can be affected by vanishing gradients, if the discriminator is too good. Also, an optimal discriminator may not present sufficient details to improve the performance of the generator. As the solutions, different loss functions such as Wasserstein loss, which addresses vanishing gradient problems during the training of discriminator, or modified max loss to deal with the same problem, can be used.

When the generator begins to generate the same output continually, the discriminator will learn to deny the output. However, there can be scenarios where the next output of the discriminator is stuck in a local minimum without progressing towards

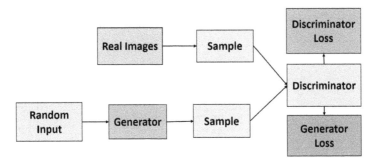

FIGURE 4.5 GAN architecture.

the optimal solution. In this case, the generator finds it easier to produce the most acceptable output in the next iteration. Accordingly, the generator overoptimizes for a given discriminator in each iteration, hence the discriminator does not learn to overcome this. Consequently, the generator rotates over a small set of output types and GAN starts to fail. This can be addressed by using Wasserstein loss and unrolled GANs to utilize a generator loss function that incorporates both discriminator's classification and also future discriminator values, where the generator cannot overoptimize its outputs for a single discriminator.

4.4 VARIATIONS OF CNNS

4.4.1 Residual Networks (ResNet)

Recently, most of the complex problems are solved by adding some additional layers in neural networks to increase model performance. The key idea behind this approach is to utilize these additional layers to learn more complex features progressively. Therefore, the deeper the architecture becomes, it will solve more complex problems with improved performance. However, when adding more layers, the accuracy of the model will be saturated and then degrade gradually. The reason for this could be the optimization function, initialization of the network, or vanishing gradient problems. The residual network (ResNet) model, which is made up of residual blocks, addresses these training issues. Figure 4.6 shows an instance of a residual block and Figure 4.7 shows the architecture of the ResNet50 model. Generally, ResNet performs well on image recognition and tasks associated with localization. The ResNet50 model consists of 50 layers and over 23 million trainable parameters. This is a fast-performing model and can train many layers without increasing the training error percentage.

As the core concept, the residual block uses a skip connection, which is a direct link that avoids some middle layers. In a general neural network without a skip connection, the input value (x) will get multiplied by the weights of the layers and add

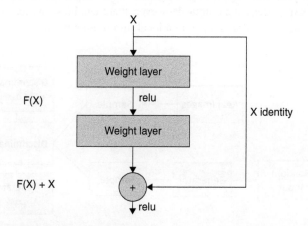

FIGURE 4.6   Residual block architecture.

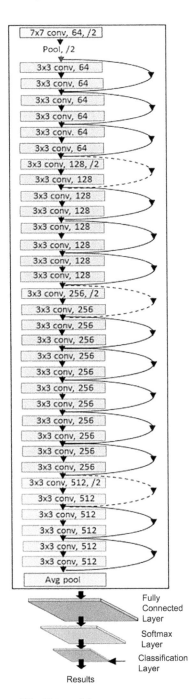

FIGURE 4.7 Architecture of ResNet model.

a bias term as in (4.1), where $f()$ is the activation function and the $H(x)$ is the output. When the skip connection is present, the output will be changed as in (4.2). However, the input dimensionality might differ from the output, which mainly happens with convolutional layers and pooling layers. This can be addressed by adding a 1 * 1 convolutional layer to the input in such a way that it fit the dimensions as in (4.3), where $w1$ denotes the additional parameters.

$$H(x) = f(wx + b) \text{ or } H(x) = f(x) \tag{4.1}$$

$$H(x) = f(x) + x \tag{4.2}$$

$$H(x) = f(x) + w1.x \tag{4.3}$$

Let us see, how the ResNet solves the gradient vanishing problem. ResNet allows the deep neural network to use alternative shortcut paths for the gradient to flow through. These skip connections in residual blocks help to learn the functions, ensuring that both the higher and lower layers will perform well. Thus, the ResNet model handles the vanishing gradient problem using identity mapping. It includes skip connections that serve as gradient superhighways, allowing the gradient to flow freely. It allows gradients to spread to deeper layers before becoming attenuated to tiny or zero levels. Another issue in model training with optimization is the use of wide parameter space. This can result in naively adding layers and increasing training loss.

4.4.2 INCEPTION MODEL

We expect to have neural networks perform well with large-scale and multi-scale convolutional layers. The model learning is based on the Hebbian principle, where the neurons activate, connect with other neurons and create a neural network to learn a new aspect. Accordingly, Inception models believe that for neural networks to be highly performant, they should scale well from all aspects.

The Inception-v3 model was introduced by Google in 2015 with 42 layers. It is a commonly applied CNN model mainly for image classification. This model provides high-performance gain on CNNs with a lower error rate. It utilizes the resources efficiently with minimal growth in computational load. Also, it supports feature extraction at varying scales with various sizes of convolutional filters. It is capable of handling the vanishing gradient problem.

An Inception model is a DNN with repeating components as shown in Figure 4.8. Here, for the dimension reduction of the data, a 1×1 convolution is used. It enables the expansion of the depth and width of the network. The 3×3 and 5×5 convolutions that lead to different convolutional filter sizes, enable the model to learn spatial patterns at various scales across all the dimensions of the depth and width of the input. The Inception model architecture is shown in Figure 4.9, and the benefits of the Inception model can be listed as follows.

- Provides good performance improvement on CNNs.
- Consumes computation resources efficiently with less computation load.

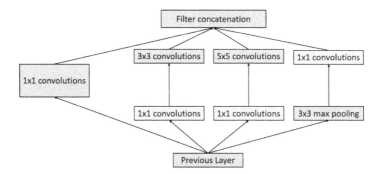

FIGURE 4.8 Inception architecture.

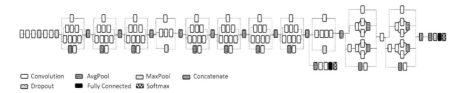

FIGURE 4.9 Architecture of the Inception model.

- Extracts input features in various scales with different sizes of convolutional filters.
- The patterns in cross-channels are learnt by 1×1 convolution filters supporting feature extractions.

4.4.3 GOOGLENET

GoogLeNet is a CNN that contains 22 layers of deep convolutional, pooling layers supported as a modification of the Inception model. It is a network with parallel concatenations and is widely used for object classifications. The importance of this architecture is that it is performing on par with the state-of-the-art models with a higher degree of computational complexity while maintaining the computational budget at a constant level. In real practice, having a model to be used in image recognition or object detection tasks should be efficient and should be able to embed in lower resources devices, such as mobile phones to get the actual usage of the application. We have learnt that the increase in the number of layers in a neural network often leads to high-performance gain, however, it is computationally intensive. More importantly, large networks become overfitting and face either a vanishing gradient problem or exploding gradient problem. The GoogLeNet model addresses the challenges of large networks, by utilizing the Inception module to increase computational efficiency. As shown in Figure 4.10, this architecture has 22 layers, which extends to 27 layers including pooling layers. A section of these layers is complied with nine Inception modules.

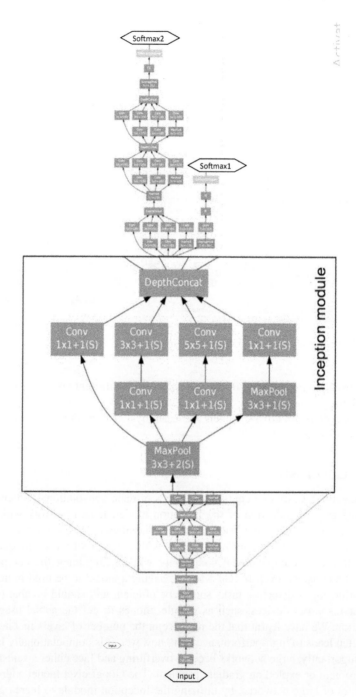

FIGURE 4.10   GoogleNet architecture.

The GoogleNet architecture alternates convolutional and max-pooling layers. The first convolutional layer utilizes a relatively large filter size of 7×7. This helps to dimensionality reduction of the input, without losing spatial details by using large filter sizes. Here, the width and height of the images are decreased by a factor of four at the second convolutional layer. Subsequently, it is reduced by a factor of eight before passing to the first Inception module. However, it generates a larger number of feature maps. The final prediction component of the GoogleNet consists of an average pooling layer, a dropout layer of 40%, and both linear and Softmax classifiers. Generally, the average pooling layer calculates the average of the feature maps generated by the last Inception network and reduces the input size to 1×1. Then the dropout layer is utilized before the linear layer for regularization to address data overfitting.

4.4.4 XCEPTION MODEL

The Xception model is an efficient architecture in terms of computational time, as it involves depthwise separable convolutions. This is an extension of the Inception architecture, where the Inception model is substituted with depthwise separable convolutions, which is followed by a pointwise convolution. Figure 4.11 shows the architecture of an Xception model, where the plus sign denotes the elementwise addition. However, it is expensive to train.

4.4.5 DENSENET MODEL

The DenseNet architecture consists of dense connections between layers as shown in Figure 4.12. Here, the dense blocks directly connect each of the layers with the same size of feature maps. This feature addresses the vanishing gradient issue in models with high depth and improves the declined accuracy. As shown in Figure 4.12, each layer acquires additional inputs from all the prior layers and sends their feature maps to all successive layers. Here, c denotes the channelwise concatenation.

One of the advantages of DenseNet is its powerful gradient flow. Here, the loss is transmitted to the previous layers easily and directly, where the previous layers obtain direct supervision from the final layer implicitly. In a DenseNet, each layer gets a combined knowledge of feature maps from all the previous layers. This may cause the network to be narrower and denser. This model provides efficient computations and memory utilization as it can have few channels and parameters. Additionally, it

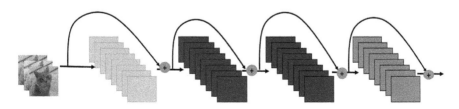

FIGURE 4.11 Xception model architecture.

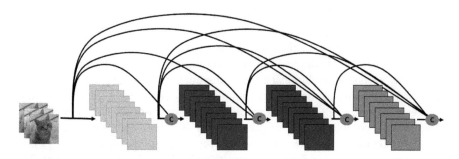

FIGURE 4.12   DenseNet model architecture.

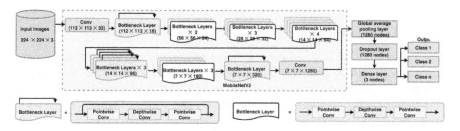

FIGURE 4.13   MobileNet model architecture.

supports expanded features and complex patterns of data, as each layer gets all the prior layers as inputs. Consequently, DenseNet provides smooth decision boundaries. Hence, supports high-performance values even with a lesser amount of input data. Although there are many advantages, DenseNet requires heavy GPU memory due to concatenation operations.

4.4.6 MOBILENET MODEL

MobileNet is a simple, end-to-end transparent pipeline architecture. It provides a lightweight and fast neural network with few parameters and uses depthwise separable convolutions as shown in Figure 4.13. Additionally, the number of parameters can be further lessened by using dense blocks. Generally, a filter in a convolutional layer is applied for all the channels of the input. This is done by taking the weighted sum of the pixels in the input image with the filter. Then it passes to the next input pixels over the image. A MobileNet model uses this convolution only in the first layer. The next set of layers is the depthwise separable convolutions consisting of both depthwise and pointwise convolutions. The depthwise convolution applies convolution for each channel of the input separately to filter the input channels. The pointwise convolution uses a 1×1 filter to merge the output channels from the depthwise convolution and produce new features. Since this requires fewer computations, these models are best suited for embedded and mobile devices.

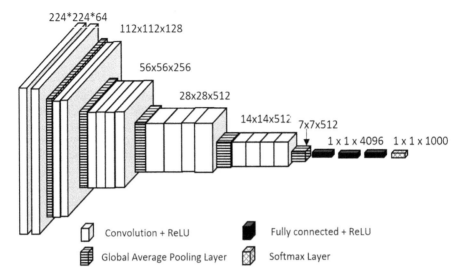

224*224*64
112x112x128
56x56x256
28x28x512
14x14x512
7x7x512
1 x 1 x 4096   1 x 1 x 1000

☐ Convolution + ReLU   ◼ Fully connected + ReLU

▤ Global Average Pooling Layer   ▨ Softmax Layer

FIGURE 4.14   VGG architecture.

4.4.7 VGG MODEL

Visual geometric group (VGG) is an innovative object recognition model based on CNN with multiple layers. VGG is a deep NN and has been trained on large datasets with different images with complex classification tasks. Thereby it is one of the widely used vision models. Based on the deepness of the model, there are VGG-16 and VGG-19 with 16 and 19 convolutional layers, respectively. As shown in Figure 4.14, it uses a set of convolution and max pool layers constantly. There are two fully connected layers and a Softmax layer before the output. The VGG model focuses on how the depth of the CNN affects the accuracy of image classification tasks and how to reduce the number of parameters. Therefore, VGG uses a small receptive field such that, 3 * 3 convolutional kernels in all its layers. There are also 1 * 1 convolutional filters as linear transformers of inputs, which are followed by ReLU activations.

4.4.8 COMPARISON OF CNN ARCHITECTURES

Let us see a comparison of a few widely used architectures. Table 4.1 shows a summary of ResNet, Inception, and VGG networks.

In earlier models, sigmoid and tanh activation functions were widely used. In order to overcome the vanishing gradient problem associated with these activation functions, the ReLU activation function has been used to train later models. Although ReLU addresses the vanishing gradient issue, it results in very high learning variables because of its unbounded nature. This can be addressed by applying normalization in a neighborhood of pixels by boosting the current neuron while contracting the nearby neurons.

TABLE 4.1

Comparison of ResNet, Inception, and VGG Networks

Model	Salient Feature	Architecture
ResNet	Shortcut connections	Has more parameters (about 60M) and requires high computations, more training time and energy. It consists of convolutional layers with 3×3 filters. At the start and the end of the model, there are two pooling layers. Identity connections exist between every two convolutional layers.
Inception	Wider–parallel kernels	A 5×5 and 3×3 convolutional layers are used to capture global and distributed features, respectively. The low-level features are identified by max-pooling. All the features are obtained and combined in each layer, and then pass to the next layer.
VGG Net	Fixed-size kernels	Has more parameters (138M) and consumes more training time. Uses 3×3 convolutional filters and 2×2 max-pooling with a stride of two. Focus on reducing the parameters in the convolutional layers and improving training time.

The selection of a neural network and hyperparameter tuning should be done with care. When we add more layers to the model and increase the depth of the network, it may reduce the accuracy due to the vanishing gradient problem. Here, the derivatives become insignificant during the backpropagation. In addition, we use dropout layers to address data overfitting. It drops connections at a specified rate during training to avoid local minima. However, it doubles the needed iterations for convergence.

In general image classification, there can be varying sizes of salient features in an image. In that case, we may not be able to use a fixed kernel (filter) size. We use large filter sizes when there are more global features are distrusted over a large region of the image. In contrast, small filter sizes perform well to identify area-specific features over the image. Therefore, different filter sizes in the same layer are used to identify variable-sized features as in the Inception model. This process considers the width of the network, instead of the depth.

Although CNNs are used in a wide variety of image classification tasks, it contains a few drawbacks that are crucial in computer vision problems. For instance, CNNs can precisely identify image edges using the initial layers and complex features using deeper layers. However, there is a limitation in identifying the spatial composition of the identified features using a CNN. Therefore, in tasks where the spatial composition is important in classification, the CNN models may not give the best results. In such situations, the pooling function that connects the layers become inefficient in detecting relationships between the object parts. Here, the pooling loses connectivity information, as the pools do not overlap.

Consider the following example of identifying a face. If we use a CNN to recreate the same image, it may output a representation as in Figure 4.15, which has not considered the positional or instantiation information of the image. In general, a

FIGURE 4.15 Face image reconstruction by a CNN.

CNN tries to identify the features in the image, but not their relative position. This is due to the feature extraction using max-pooling without overlapping areas. Therefore, it is important to use an approach, where the pools are overlapped to preserve the positions of features.

4.5 CAPSULE NETWORK

A capsule network (CapsNet) is an extension of an ANN that supports hierarchical connections and preserves spatial composition. CapsNet overcomes the loss of information that is seen in pooling operations, hence fetching more important features. A capsule gives a vector as an output that has a direction. The probability of an existence of a given instance is given by the length of the vector. The instantiation parameters, such as position, orientation, scale, and color, are given by the orientation vector. Capsules are the functional elements that perform inverse rendering, which is the extraction of instantiation parameters. Therefore, a capsule predicts the existence and the instantiation parameters of an object in a given location.

The standard neural networks use neurons to extract features, while capsule models capture only the crucial image details. Compared to general neurons that generate scalar quantities, the capsules generate vectors that can recognize the direction of a given feature. When the direction of a given feature is changed, only the direction changed corresponds to the position change, but the value of the vector remains the same. These models provide good results with small datasets as well, by easily interpreting the robustness of the images.

The term 'routing-by-agreement' (dynamic routing) explains the routing of capsules between two consecutive layers and addresses the inefficiency in max-pooling due to information loss. Here, the capsules of the first layer predict the output of the second layer. Consider an example of an image of a person. Here, only the lower-level features, such as mouth and eyes with matching content, are sent to the corresponding higher level. Here, the feature representing the eyes and mouth will be sent to the layer representing the features of a face and the features, such as fingers, nails, and palms, will be sent to the corresponding layer of a hand. Thus, the process encodes spatial details into the features with dynamic routing.

Therefore, a capsule network learns the features by reproducing the image using the extracted features. It learns the prediction of the output, by reproducing the expected object and assessing it against a labeled instance in the training set. Since the generated predictions will be selected based on the highest expected probability of most of the capsules, it sends a clean input to the next layer and requires fewer capsule layers to service perfect results. This in turn reduces the training time and resources. Also, by navigating backwards of the network the possession of the elements in the object can be precisely derived, hence image reconstruction is feasible. This process leads to predicting better instantiation parameters.

The dynamic routing function acts as a dual loss function to get better predictions. It moves the feature around the image, calculates the probability of each of the positions and locates the feature in the most appropriate location. This is known as equivariance between capsules. In a capsule network, the process starts by calculating the dot product of the weight matrix and the input vectors by the capsule. During the routing process, the capsules in the lower-level pass the inputs to the capsules in the higher level by encoding the spatial relationship among the layers. Then dynamic routing is used to select a parent capsule for a given capsule, by ensuring that the output of a given capsule goes to an appropriate parent in the above layer. Here, a function is used to squash a vector to a value between 0 to 1, while keeping the same direction.

The architecture of a capsule network consists of six layers comprising an encoder and a decoder. The first three layers correspond to the encoder and they transform the input into a 16-dimensional vector.

As shown in Figure 4.16 these layers can be explained as follows.

1. The first layer of the encoder is a Conv layer, and it extracts the basic features of an image that can later be analyzed by the capsules. It has 256 kernels of size 9×9×1.
2. The second layer is the PrimaryCaps Network, which consists of a different number of capsules, takes essential features, and finds more detailed patterns

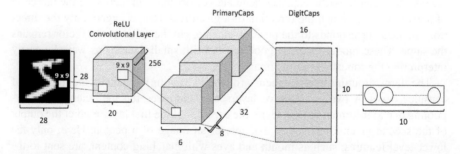

FIGURE 4.16 Capsule network architecture: encoder.

with spatial relationships. This is the lower-level capsule layer that contains 32 capsules. Each of these capsules utilizes eighth 9×9×256 convolutional filters to the result of the prior convolutional layer and produces a 4D vector output.

3. The third layer is the DigitCaps Network, which contains different capsules. This is a high-level layer, where the primary capsules are transmitted in dynamic routing. After these layers, the encoder has a 16-dimensional vector with the required details to render the image that goes to the decoder.

Accordingly, the capsules generate the positional and instantiation information by a routing mechanism and compute the marginal loss to decide the best instantiation representation and then that information is passed to a decoder network to reconstruct the image. The decoder is a feed-forward network with three fully connected layers as shown in Figure 4.17. Once the encoder capsule with a 16-dimensional vector passes to the decoder capsule, it reconstructs the same image from scratch with its data. The decoder learns to decode the instantiation parameters of the input image and uses the Euclidean distance loss function to assess the similarity between the actual and predicted outputs. As mentioned, since the capsules keep only the significant details to identify the features in the vector, it produces robust results.

Generally, capsule networks provide high results with small and simple datasets such as MNIST. However, it is hard to apply for complex and advanced datasets such as CIFAR-10 or Imagenet, due to the extra details being difficult to handle by the capsules. The advantages and disadvantages of the capsule network are listed as follows.

Advantages of capsule networks:

1. Work with small training datasets.
2. Identify hierarchical feature representation of data.
3. Preserve positional parameters of features.
4. Use interpretable activation vectors.
5. Consider overlapping objects using routing by agreement.

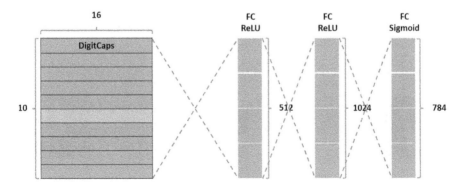

FIGURE 4.17 Capsule network architecture: decoder.

Disadvantages of capsule networks:

1. Low performance with complex or large datasets.
2. The training process is slow because of the inner loops.
3. Hard to differentiate identical objects of the same type that are positioned close to each other.

4.6 AUTOENCODERS

An autoencoder is an extension of ANN that efficiently learns data representation (coding) of unlabeled data through unsupervised learning by ignoring the noise signals. Generally, an autoencoder is a dimensionality reduction feedforward neural network where the input is the same as the output. It is mainly used for denoising, compression, and data generation of images. As shown in Figure 4.18, it has three modules namely encoder, code, and decoder. The encoder, which is a fully connected ANN, squeezes the input to generate a latent-space representation, which is a dimensionality reduction code. The decoder uses this compressed code to rebuild the input by refining and validating the encoded content. Figure 4.19 shows the architectural visualization of an autoencoder. The code is a single layer and the size of the code, that is the number of nodes in the code layer, is a hyperparameter that needs to be defined before the training process. The bottleneck architecture restricts the amount of data that traverses through the network and forces it to prioritize the features of the input that should be copied. The output is closely similar to the input, where both dimensionalities are the same.

The types of four hyperparameters to be defined before the training process:

•   The number of layers: the depth of the model can vary based on the problem. In Figure 4.19, both the encoder and decoder have two layers, except the input layer and output layer.
•   The number of nodes stacked in each layer of encode and decoder: in the encoder, the number of nodes in a layer reduces when moving with each subsequent layer. In the decoder, the number of nodes in the subsequent layers increases correspondingly. Therefore, the layer structure in the encoder and decoder is symmetric.
•   Code size: indicates the number of nodes in the middle layer (code). More compression is provided when the code size is small.

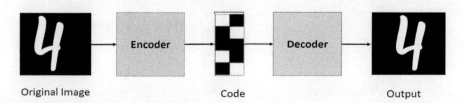

Original Image   Code   Output

FIGURE 4.18   Process flow of autoencoder.

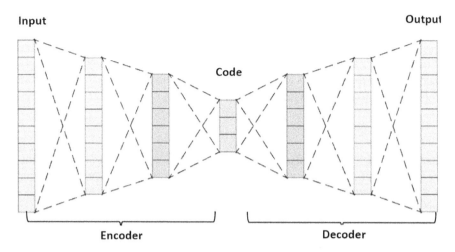

FIGURE 4.19 Autoencoder visualization.

- Loss function: uses cross-entropy loss when the input values are between zero and one. If not, the mean squared error is used.

The properties of autoencoders are listed as follows.

- Data-specific: autoencoders can squeeze data, that are similar to the trained data.
- Self-supervised: it does not need explicit labels to train, and generates labels from the training data.
- Lossy: the output is a reduced representation of the input.

Let us learn a few standard autoencoder architectures.

- Complete autoencoders

Autoencoder trains to perform the task of copying original data taking only the useful features. In complete autoencoders, the code layer's dimension is lower than the dimension of the input. The learning process of an autoencoder tries to minimize its loss function. This can learn the most important features from input data and efficiently reconstruct the original data by penalizing the model based on the reconstruction error.

- Regularized autoencoders

The encoders and decoders with more capacity may fail to learn useful information in the data. A similar problem may occur when the hidden code allows dimensions equal to the input. Also in the case of overcomplete, where the hidden code contains dimensions larger than the dimensions of the input, the linear encoder and decoder will learn to replicate the same input as the output, instead of learning effective features.

Therefore, it is important to select the dimension of the code and the capacity of the encoder and decoder based on the data complexity. The regularized autoencoders utilize a loss function to avoid copying input data to its output during learning, instead of reducing the capacity of the model by maintaining a small and shallow encoder and decoder. These autoencoders can be nonlinear and overcomplete, but they will always learn important information from the data.

- Sparse autoencoders

An autoencoder can be regularized by using a sparsity constraint. In this case, only a set of nodes will become active nodes by having non-zero. This represents the input as a small subset of active nodes and learns to identify important features. Thus, it performs well even with large code sizes. It uses sparsity to create an information bottleneck. However, for natural image data, regularized autoencoders and sparse coding tend to yield very similarly.

- Denoising autoencoders

Denoising autoencoders learn useful features by modifying the loss value, rather than applying a penalty as in sparse autoencoders and having a small code layer as in complete autoencoders. Denoising autoencoders add random Gaussian noise to their original images. These data with noise are fed as input to the autoencoder and it regenerates output without noise. Since the input consists of some noise, the autoencoder cannot directly reproduce the output from the corresponding input. Thus, these denoising autoencoders subtract the noise and produce effective data as shown in Figure 4.20. The architecture of a denoising autoencoder is similar to a regular autoencoder, except for the fit function. Denoising autoencoders use the function $L(x, g(f(\hat{x})))$, where x is the original input and $\hat{x}$ is the input with noise. The process tries to handle the noise. Now let us see the usefulness of autoencoders at compressing the input. Although, they are not commonly used in real applications because of their data-specific nature. Therefore, they are used as a preprocessing step for dimensionality reduction.

4.7 TRANSFORMERS

Transformers are used to transform an input sequence into an output sequence. This process is known as sequence transduction. Transformers often apply to

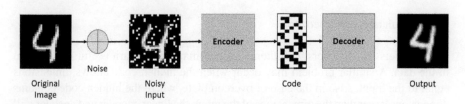

FIGURE 4.20 Denoising autoencoder.

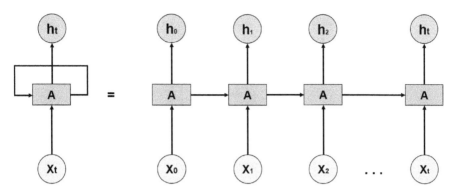

FIGURE 4.21   Recurrent neural network architecture.

real-life applications such as speech recognition and other sequential data processing applications like text-to-speech transformation. This requires understanding
the dependencies and connections within the context referred to. Thus, transformers
utilize a self-attention mechanism that applies different weights for specific regions
of the input. The transformer model is designed in such that it trains parallelly both
data and model. Thus, these models are more efficient than RNNs such as long-short
term memory (LSTM). The effect and efficiency are balanced using an encoder-
decoder architecture.

First, let us refresh your understanding of RNN and LSTM. As we have learned,
RNNs have loops in them that act as memory as shown in Figure 4.21. RNN creates
and connects several copies of the same model to provide the prediction. Therefore,
the sequence data are handled by a chain-like nature. Consider an application that
translates a set of words, where a word in the text is a given input. The RNN sends the
details of the prior words to the subsequent model for further processing. Generally,
RNNs do not perform well when there is a large gap between the relevant knowledge
and the location where it is used. Also, since the details are passed in each model, the
chain becomes longer allowing information loss.

If we extract the importance of the information presented in the form of data,
RNNs try to completely convert the data into new forms of information by applying
a function. Therefore, the importance of the information may not be persisted. Also,
RNNs have gradient issues. To address this LSTM architecture has been introduced. It
is a variety of RNNs that learns long-term dependencies, which is required for the prediction of data sequences. The feedback connections of the LSTM support processing
the complete data sequence, instead of a single data location as in images. LSTM
tries to modify the data by performing a few operations, such as multiplications and
additions, the information flow mechanism of LSTM is known as cell states, where
LSTM selectively learn or forgets information that is crucial or not. Figure 4.22
shows the architecture of an LSTM.

Figure 4.23 shows a comparison of RNN and LSTM. We explained the architecture of RNNs, which consists of a sequence of repeating neural network modules..
In the original RNN architecture (left figure), these duplicating modules represent a

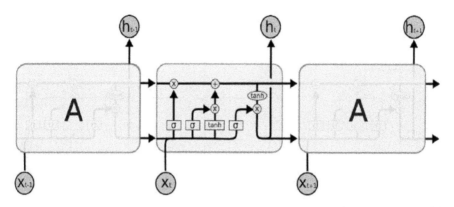

FIGURE 4.22 LSTM architecture.

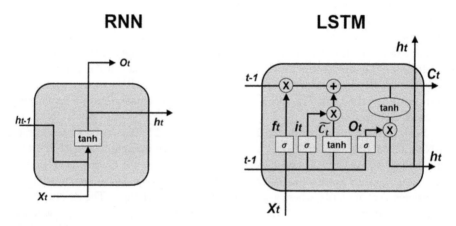

FIGURE 4.23 Comparison of RNN and LSTM.

simple structure as indicated by the tanh layer. In contrast, in the LSTM architecture (right figure) each duplicate module has a different structure with four neural network layers.

The same challenge of handling longer sentences faced by RNNs applies to LSTMs as well. When there are long sentences, the performance degrades with the increase of the distance between the word in the context and the processing point of that word. Another issue in both types of models is since it processes word-by-word it is difficult to process sentences parallelly.

Additionally, the long- and short-range dependencies are not considered by these models. This can be addressed by the concept of attention that considers the linear distance among the positions and the dependencies in long and short ranges. In this mechanism, special attention is given to the word of the current translation when the translation is happening. In a neural network, this can be applied in several ways.

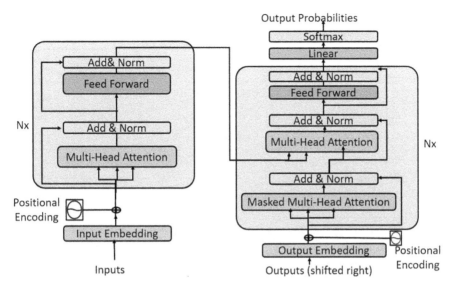

FIGURE 4.24 Transformer architecture.

For instance, in an RNN, there is a hidden state for each word, which is sent to the decoder, without encoding the complete sentence in a hidden state.

In CNN, we address a couple of these problems with inherent features. For instance, we can parallelize the translation process, the distance between the positions becomes logarithmic and the local dependencies can be exploited. However, CNNs do not handle the dependency issues in sentence translations. In order to address these issues, a combination of CNN and the new concept of attention has been used in the transformers. Figure 4.24 shows a transformer architecture consisting of an encoder and a decoder. These components are compiled in such a way that they can be created by stacking on top of each other multiple times to support the chain architecture introduced in RNN, as shown in Figure 4.25.

The encoder has a feed-forward neural network and a self-attention layer or multi-head attention layer. The attention technique draws connections between any parts of the sequence and addresses the issues with long-range dependencies. With transformers, long-range dependencies have the same likelihood of being considered as any other short-range dependencies. The attention technique addresses this dependency by refer-back to the hidden state of the encoder by the decoder by its current state. Informally, this can be considered as a variation of the encoder-decoder model with bi-directional LSTM. Therefore, the decoder can retrieve only the crucial details of the input and learn complex dependencies between the input and the output.

Attention is calculated by dividing the inner product of the sender (query) and receiver (key or target word of the attention), by the square root of its dimensions. Each word creates its attention for all the other words by giving a query to comply with a key. When a sentence consists of many words, since there are more inner products, the square root is considered as a variance balance. By applying the

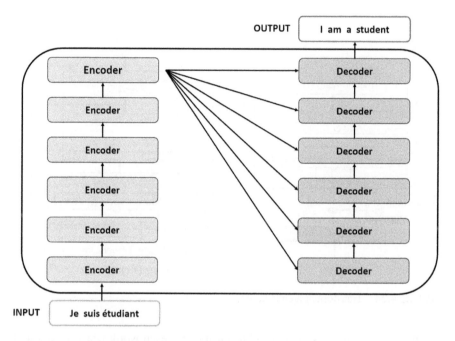

FIGURE 4.25   Stacked architecture of the transformer.

Softmax function through the attention matrix, the normalized matrix can be derived. The feed-forward network is utilized for each attention vector or matrix that is created to convert the attention vectors into a structure that can be fed into the next layer of decoding or the next encoder layer. The decoder network also consists of a stack of multi-head attention, feed-forward networks, and an ending system with linear or Softmax classifiers like CNN architecture. Additionally, the decoder block has a masked (hidden) multi-head attention block.

Vision transformer (ViT) is used for vision processing tasks. It has the nature of transformers that are used for natural language processing, where the learning is based on the relationship between input token pairs. Generally, this lacks the inductive biases of CNNs, such as locally restricted receptive fields and translation invariance. That is, it identifies an object in an image, with varying appearances or positions. Recall, that convolution is a linear local operator that observes the neighboring values that are specified by the kernel.

The transformer is permutation invariant and cannot process grid-structured data. Thus, the need to transform a spatial non-sequential signal into a sequence. Transformers outperform CNNs in accuracy and computational efficiency. The term translation indicates that each pixel of the image moves in a given direction with a fixed value. Overall, the visual transformer initially splits an image into fixed-size patches and embeds them. Then, as shown in Figure 4.26, it incorporates positional embedding as an input to the transformer encoder. It uses multi-head self-to-remove image-specific inductive biases to understand the image features.

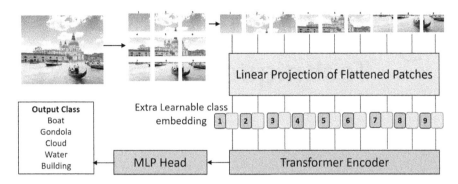

FIGURE 4.26   Process of visual transformer.

The process of a visual transformer can be listed as follows.

1. Split an image into patches.
2. Flatten the patches.
3. Use the flattened patches to generate lower-dimensional linear embeddings.
4. Add positional embeddings.
5. Send the sequence to a standard transformer encoder.
6. Pretrain the model with image labels (supervised on a large dataset).
7. Classify the image by refining the downstream dataset.

REVIEW QUESTIONS

1. Compare and contrast the following architectures.
   a. Multilayer perceptrons and convolutional neural network.
   b. Feed-forward neural network and multilayer perceptrons.
   c. Recurrent neural network and convolutional neural network.
2. Explain potential applications of capsule neural networks and their ability to regenerate data.
3. Explain why autoencoders are the most popular neural network element and their usage.
4. Explain why multilayer architectures are becoming popular and their potential drawbacks.
5. What is significant about the parallel concatenation of the networks and describe its advantages?

5 Advanced Learning Techniques

5.1 TRANSFER LEARNING

5.1.1 Overview of Transfer Learning

Humans can learn a task and transfer the obtained knowledge to resolve associated tasks using their inherent nature. The concept of transfer learning utilizes the learned knowledge about a task to solve related problems, without relying on solitary learning. Transfer learning takes a model, which is trained on the same domain, with different tasks or different domains with the same task. Subsequently, the learning process adapts to the considered domain and the target task, without training the model from scratch.

Transfer learning takes a model trained on a large dataset and transfers the obtained intelligence to another dataset. For instance, consider the task of recognizing an object in an image using a CNN. Following the transfer learning approach, the initial set of convolutional layers of the considered pretrained model can be frozen and train only the set of layers in the latter part of the model to predict the results. Generally, the initial convolutional layers provide low-level feature extraction over the image, such as edges, gradients, and patterns. The last set of convolutional layers supports extracting complex features that will lead to recognizing objects correctly. Accordingly, a pretrained model on an unrelated large dataset can be utilized to predict a task, because the general low-level features are common to many images, as shown in Figure 5.1.

As shown in Figure 5.2, we can distinguish transfer learning from traditional machine learning. Traditional learning algorithms start from scratch and depend only on the considered dataset and the allocated application. They do not keep the learned intelligence to be used by another model. Here, the learning is performed without comparing the knowledge that is learned in the past with the new tasks. In contrast, transfer learning learns new responsibilities that rely on the tasks that were learnt before. It leverages the associated weights and features, weights in the pretrained models to train new models by generalizing the knowledge. Importantly, this can handle the issues of insufficient datasets for new tasks. Consequently, this learning process can be fast and more accurate.

DOI: 10.1201/9781003390824-5

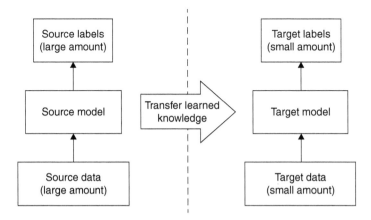

FIGURE 5.1 Overview of transfer learning.

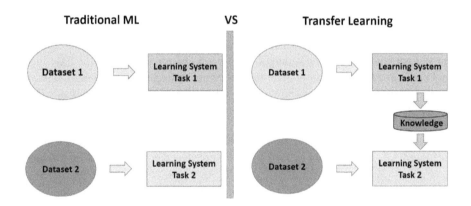

FIGURE 5.2 Traditional ML vs transfer learning.

5.1.2 TRANSFER LEARNING PROCESS

The general process of transfer learning for object recognition can be listed as follows.

- Use a CNN model, which is pretrained on a large dataset.
- Create a base model.
- Freeze the weight parameters in the first few convolutional layers of the model. The adjustment of the freezing layers is decided based on the compliance of the original dataset with the new task.
- Substitute the last set of layers of the model with a modified classifier with several layers of trainable parameters of the model. For instance, some of the custom classifiers could be a fully connected layer with ReLU activation, dropout with $x\%$ possibility of dropping and fully connected with Softmax activation.

- Set the number of outputs as the same as the number of classes.
- Train only the modified classifier layers on training data to optimize the model for a smaller dataset.
- Refine hyperparameters and try unfreezing more layers.

Accordingly, the pretrained model is loaded initially and the base model is initiated. For instance, some popular modules in computer vision are ResNet, Inception, Xception, and VGG and in natural language processing, we have Word2Vec and Glove language, models. As an option, the pretrained weights can be downloaded, or the selected model can be trained from scratch. We need to freeze a few layers from the pretrained model to avoid changing during training. Thus, the weights associated with these layers would not be reinitialized. If the weights are changed, then it will not be based on the prelearned knowledge and can be considered as a model trained from scratch.

Usually, the base model contains a different set of elements in the output layer based on the number of classes of the considered application, as the outputs of the pretrained model and the new model are different. For example, generally, the pretrained models are trained on the ImageNet dataset that output 1000 classes. However, the new model will have two or three classes. Therefore, the model should be trained with a new layer for the output. Thus, the final output layer has to be removed from the base model and needs to add a new output layer that is corresponding to the number of classes in the considered application as shown in Figure 5.3. Then, a set of new trainable layers are added to train the model with the existing features to make the predictions for the new data.

After that, the model performance can be improved through fine-tuning as shown in Figure 5.4. The features extracted by the pretrained model should be fine-tuned

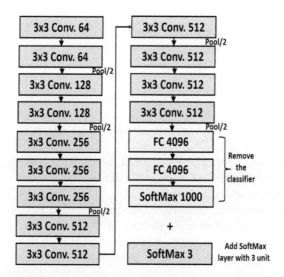

FIGURE 5.3   Replacing layers of the base model.

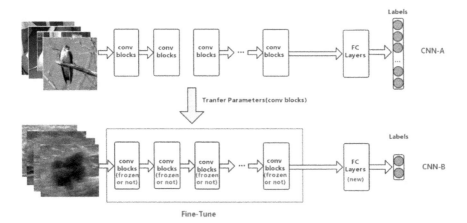

FIGURE 5.4 Fine-tune process.

to obtain the new features specific to the new base model. The fine-tuning process unfreezes the set of layers of the base model and trains the model for the entire dataset. This starts with a low learning rate as a large model is trained on a small dataset. Also, this avoids higher value changes in the gradient that can lead to low performance. Accordingly, it prevents data overfitting and increases performance.

Generally, a model training over the dataset continues to repeat for a given number of epochs reached. However, the model training could result in overfitting. As we learnt in previous chapters, early stopping helps to address this issue. It stops the training when the validation loss does not reduce or reaches a plateau state or starts to increase for a consecutive set of epochs, which only the training loss reduces. To find the generalized model for the test dataset, we compare the parameters corresponding to each epoch during the region that reduces the validation loss and identify the parameters with the best validation performance. Accordingly, when the learning curve does not show any improvement, it is necessary to hold the training.

5.1.3 TRANSFER LEARNING TYPES, CATEGORIES, AND STRATEGIES

Different types, categories, and strategies are available in transfer learning as shown in Figure 5.5. They are selected based on the learning type. For instance, multitasking learning supports learning different tasks parallelly in the same domain. Domain adaptation is a transfer learning type with distinct feature spaces and distributions that adapts several sources to improve the target.

• Transfer learning types

Different types of transfer learning algorithms exist, such as inductive, unsupervised, and transductive transfer learning. Inductive transfer learning consists of the same source and target domains, but different associated tasks. Similarly, unsupervised transfer learning has the same setting but mainly considers the unsupervised tasks

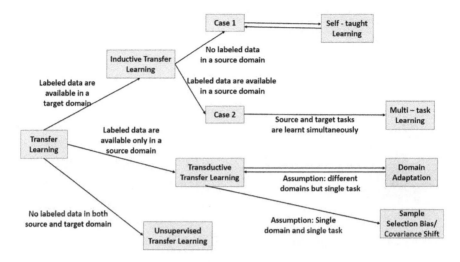

FIGURE 5.5   Transfer learning strategies.

with unlabeled data in the target domain. In transductive transfer learning, there will be some similarities between the tasks that correspond to the source and target but has different domains. There is a large set of labeled data in the source domain, while no data is available in the target domain.

- Transfer learning categories

When transferring across these different categories, there may be a question about what to transfer. This can be addressed by transferring parameters, instances, features, and relational knowledge. The instance transferring reuses a set of source instances with the target domain to enhance the result. Inductive learning uses modifications to utilize training the source domain instances to improve target tasks. In feature representation transfer, the error or domain divergence is minimized by identifying the most relevant features that can be effectively applied from source to target domains. As the name implies, the parameter transfer shares parameters among the models with related tasks. Thus, the additional weight is applied to rectify the error values in the target domain to increase the accuracy. Unlike all these transfer learnings, the relational knowledge transfer handles are mainly used on non-independent and iden-tically distributed data such as social networks.

- Transfer learning strategies

As we discussed, pretrained models support feature extraction and fine-tuning during transfer learning. Generally, neural networks consist of many hidden layers with tun-able hyperparameters such that the basic and complex features are captured by earlier and later layers, respectively. Therefore, the complex feature representations in later layers of the base model need to be fine-tuned to train those layers to extract more specific features of that task. Consequently, while keeping some of the earlier layers frozen, the transfer learning process retrains the latter layers. The final layer is a fully connected layer that connects the outputs from the previous layers. Here, the layered architecture utilizes the pretrained models by replacing the final layer compatible with the outputs of the selected task.

Another strategy is using fine-tuned, pretrained models, where the final layer is not replaced, but retrained some of the previous layers are. Using this approach, we can retrain or fine-tune selected layers to have a better performance with less time for training. As we leant, initial layers learn basic features that are generalizable to most types of data. The higher layers extract features that are more specific to the considered dataset for training. The fine-tuning process helps to utilize the specific feature representations to comply with the new dataset. Therefore, these layers can be frozen and reused with the basic knowledge derived from past training.

*   Transfer learning methods

One-shot learning and zero-shot learning are variants of transfer learning. One-shot learning gives only one labeled sample for the transfer task and no labeled samples are given for the zero-shot learning task.

One-shot learning learns from one or a small number of instances to classify many new instances to infer the required output. Generally, this method is used with insufficient labeled data or with more new classes. For instance, face recognition applications classify faces of people with varying lighting conditions, hairstyles, accessories, and expressions where the model has a smaller number of images as input. Thus, it is based on the knowledge obtained by training the base model with a small amount of data per class.

Zero-shot learning based on unlabeled training data. It makes alterations during the model training to generate extra details for the unseen data. This concept of transfer learning is mainly applied to machine translation with NLP with unlabeled data in the target language, computer vision, and speech recognition tasks.

5.1.4   TRANSFER LEARNING APPLICATIONS

Transfer learning supports increasing the performances of models for several tasks and domains that can easily get the knowledge from another model, tasks with insufficient or unlabeled data and complex tasks with constraints. The following are widely used application domains.

*   Computer vision

Most of the image classification tasks with complex feature representations are performed using neural networks. Mainly the fully connected layer recognizes the image objects together with the fine-tuning of the latter set of layers. Since image analysis tasks can be enhanced by the learned knowledge and identified patterns in similar images, the transfer learning approaches are widely used for object detection, image classification and captioning tasks.

- Audio processing

Transfer learning algorithms are used to solve tasks such as acoustic recognition or translation from speech to text. For example, when the model trained for English speech classification is running at the backend, it can be used as a pretrained model for the French speech classification model.

- Natural language processing (NLP)

The pretrained models such as BERT (bi-directional encoder representations from transformers), Albert, and XLNet can be used for linguistic processing in cross-domain tasks such as prediction of the next word and questioning-answering tasks.

5.1.5 Transfer Learning Challenges

The challenges in transfer learning that degrades the performance can be explained as follows.

- Negative transfer: the knowledge transferring from the pretrained model to the new base model results in reduced performance. This can happen when the source and target are completely unrelated or when the transferring process has a lack of influence on the relationship between the two tasks.
- Drift: the environment changes can affect the relationship between the source and target tasks. This will reduce the performance of the model.
- Transfer bounds: the quality and practicability of the transferring process can be improved by quantifying the transfer.

Generally, high performance cannot be obtained by the transfer learning process due to several reasons. For instance, when the features learnt by the initial layers are failed to distinguish the output classes, performance degradation happens. Consider an image classification task that identifies whether a door is opened or closed. The pretrained model will support detection of whether there is a door in an image. However, it may not be able to classify whether the door is opened or closed. In such scenarios, the initial set of layers needs to be retrained to extract the needed feature representation. Another aspect of low performance occurs due to the removal of layers from the pretrained model as it lessens the trainable parameters and leads to overfitting. Therefore, identifying the number of removable layers and avoiding overfitting is a process that consumes more time and effort. Further, when the datasets of the source and the target tasks are not related, the corresponding feature-transferring process will not perform well. Considering the above-mentioned aspects, the initialization of the pretrained weights can lead to better predictions compared to the use of random weight.

5.2   REINFORCEMENT LEARNING

5.2.1   Overview of Reinforcement Learning

Reinforcement learning is a major approach in machine learning that uses the concept of intelligent agents to obtain optimal results in the considered domain. Its decision-making process observes the environment and selects events from the action space to maximize the rewards over time. Here, a potentially complex environment with uncertainty is tuned into real-world scenarios, where the machine learning model would employ experiments to find possible solutions. The computational model would then act as an agent to support decision-making, where the agent receives rewards or penalties for the corresponding actions. The objective is to achieve the highest number of total rewards as shown in Figure 5.6. A detailed explanation is given later in this chapter.

Let us see how reinforcement learning distinguishes it from traditional machine learning as shown in Figure 5.7.

Deep learning and reinforcement learning are not mutually exclusive. Thus, there is no strict division among these approaches, where both use computationally created rules for autonomous problem-solving. Deep reinforcement learning is a specialization of deep learning. Considering the difference, deep learning users the training dataset to learn the model and applies the learnt knowledge to an unseen related dataset to predict results. Reinforcement learning follows a dynamic process to learn the model and adjust the model learning based on continuous feedback to obtain better predictions. Thus, compared to supervised learning, which maps functions from input to output, reinforcement learning is based on input and feedback.

5.2.2   Reinforcement Learning Process

The basic elements of a reinforcement learning algorithm are agent, environment, and reward. An agent is a fully automated or partially automated intelligent model based on deep learning. Among different agent types, human agents have sensory organs, such as the nose, and eyes and other organs, such as hands and legs. A robotic agent uses sensors, such as cameras, microphones, and infrared range finders, and

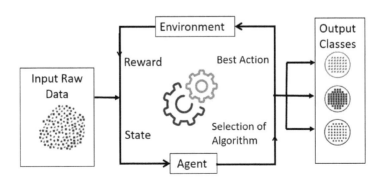

FIGURE 5.6   Overview of reinforcement learning.

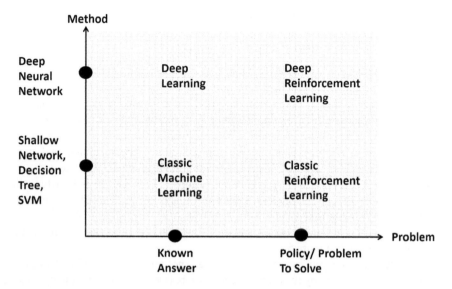

FIGURE 5.7   Traditional machine learning vs reinforcement learning.

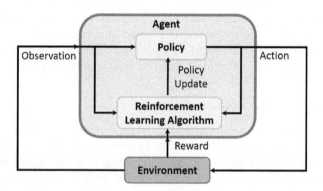

FIGURE 5.8   Process of reinforcement learning.

other actuators. A software agent consists of programs with encoded bit strings. Accordingly, deep learning applications with agents can be listed as face recognition systems, self-driving cars, chatbots, and other intelligent systems. Thus, we can define an agent as a model that uses sensors to observe the environment and uses effectors to perform activities in that environment, where the agent is located and interact. However, the agent does not control the environment with its actions.

As shown in Figure 5.8, the components of a basic reinforcement learning algorithm are listed as follows.

- Agent: the program or the model that can be trained to perform a given task. It can select an action to commit in the current state.

- Environment: the world where the agent performs his actions. It offers new inputs to the agent as a reply to the corresponding action.
- Rewards: the evaluation of an action, which can be positive or negative. This is the incentive or cumulative mechanism returned by the environment.

Reinforcement learning aims to train an agent to act in an indeterminate environment. The agent observes the environment and obtains feedback from the environment in each time interval, changes the state based on the feedback of the previous action and performs the next action back to the environment. Based on the nature of the considered environment, whether it is stochastic or not, the agent explores it to obtain the best possible feedback by changing the learning based on the feedback.

Consider a gaming application that is developed using transfer learning. The reward mechanism is set by the designer, and it defines the guidelines of the game. Initially, the model starts with no knowledge and without any suggestions or recommendations to perform the game. The model starts from random trials and for each action it performs, it will get feedback as rewards or penalties. Finally, the model learns to perform the task by maximizing the reward. This mechanism especially remodels human thinking in solving a problem. Therefore, reinforcement learning impacts effectively the intelligence of the model. Compared to human nature, this technique learns from multiple gameplays simultaneously.

Let us consider an example of playing the Alpha Go game in a real-world scenario. It is a two-user board game. In this game, a given player tries to acquire more territory than the other player. The input would be the current structural, positional information of the board and the output would be the net movement of the professional player. Even though the game rules are well-known, it is complex to decide the next action targeting the winning mode, because of the large number of moving possibilities. When humans play the game, they tackle this by calculating the impact of the movement that leads to a given position on the board. Another such application is a chess board game.

There are multiple issues associated with the classical supervised learning approach for this problem-solving. One issue would be that the ground truth of this game can be probably wrongly defined from the start. Another issue is there are many possible states and mostly this model cannot be generalized at all. Therefore, reinforcement learning can be used to provide solutions for deployed labels and make sequences of decisions.

Other than the agent and the environment, the elements of reinforcement learning can be listed as follows.

1. Action: a move made by an agent, which causes a status change in the environment.
2. State: present state sent by the environment.
3. Reward: feedback sent by the environment to assess the last action.
4. Policy: rules used to decide the next action based on the present state.
5. Value: long-term return of the present state based on the policy, instead of the short-term reward.

6. Q-value or action-value: this is an extension of value, which uses the current action as an extra parameter. It is the long-term return of the current state, for an action based on a given policy.

The policy can be improved based on the feedback obtained by the system on the value function. This is known as policy iteration. Therefore, continuous evaluation and refinement of the policy are achieved through this. A model has a set of instructions to refine the policy.

The value function identifies the feedback incentives for a given state based on a given policy. Similarly, it calculates the total rewards received by an agent from a given state based on a given policy. The value function calculates the possible value using the following methods.

- Policy evaluation: initialized using a random policy and assess the status of the state. Repeating this process, the value of the current state decides the next state that will result in the best reward. After that, the model decides the required action.
- Dynamic programming: use the reward after performing an action to calculate the value of the following state.
- Monte Carlo: execute the policy and perform the entire tasks to identify all the feedback.

However, there can be situations, where the model cannot be specifically defined. Therefore, a new function is defined as an action-value function, which calculates the predicted rewards of an action. Since this process needs to track more data, it may not perform optimal learning with deep learning. This can be addressed by using a deep Q-Network based on a neural network. The idea is to use off-policy learning. It estimates a future reward by performing an action in a given state and working towards a target policy.

Furthermore, Markov decision process is a main element in reinforcement learning. It formalized sequential decision-making where actions from a state not just influence the immediate reward but also the subsequent state. It is a very useful framework to model problems that maximize longer-term returns by taking a sequence of actions. The details of this process are not covered within the scope of this book.

5.2.3 Implementation and Scheduling Types

The focus of reinforcement learning is to identify new data points based on the previous data points. Initially, we need to find a set of actions that maximizes the predicted rewards or minimizes the cost with experiments. There are three types of reinforcement learning implementations as follows to support these goals.

- Policy-based RL uses a policy to maximize the total reward. It derives a policy directly to maximize rewards using the gradient descent method.
- Value-based RL uses value or Q-value to detect the optimal path. It tries to maximize a given value function. It analyzes the impact of reaching a given state by a given action.

- Model-based RL uses a virtual model to support agent learning in constrained environments. It uses the model to predict the next state after taking an action that has the maximum rewards.

Furthermore, reinforcement learning uses different schedulers. It is a rule that decides the reinforced behavior time of an instance. Different schedules are used to strengthen specific behaviors, as listed below.

- Fixed ratio: the feedback is rewarded after a given set of feedback instances. This is steady and leads to provide a high response rate.
- Variable ratio: the feedback is rewarded after a random set of feedback instances. Leads to a steady high response rate.
- Fixed interval: the feedback is rewarded after a predefined time interval. It shows low replying immediately after the reinforcer has occurred and shows a high replying rate closer to the end of the time interval.
- Variable interval: the feedback is rewarded after a random time interval. It shows a steady and slow response rate.

5.2.4 APPLICATIONS OF REINFORCEMENT LEARNING

Reinforcement learning is used for automated applications that perform tasks based on a set of rules to follow but not having a strict way of executing the actions. It uses computations to obtain the highest reward by trying out different actions based on the policies. A wide range of applications is implemented with reinforcement learning covering areas such as recommendation systems, robot controlling, and gaming. Following are some real-world examples of reinforcement learning.

- Learning of autonomous cars: the computer gets no instructions on driving the car. The agent learns from the rewards and penalties.
- Learning robot walk: initially, the robot takes long forward steps and fails to result in negative feedback. Thus, the model updates the learning based on the feedback and tries out a new short step that will result in rewards to allow moving forward.
- Generate user-specific advertisements: evaluate responses to messaging and discover the absolute frequency for customers. They can additionally apprise real-time bidding action in the programmatic marketplace, applying predictions regarding consumer behavior to determine which display ads to purchase.
- Computer games such as Break Out: use Google's Deep Mind Q-learning to obtain a high reward. It aims to perform an action that moves the bottom bar to bound the ball upwards in such a way that breaks the bricks.

5.2.5 CHALLENGES OF REINFORCEMENT LEARNING

Based on the application, it is challenging to relocate a model from a simulated training configuration, which is dependent on the task, to a real environment. For example, preparing a simulation environment for a model trained for a board game is easier, when it is computer-based. However, creating a simulation environment for a self-driving car is complex, as it will be implemented in the real-world considering many constraints, such as preventing collision with other vehicles afterwards.

Another challenge is based on scaling and modifying the neural network that controls the agent. Since the communication is mainly based on feedback, it may result in information loss, due to replacing the previous knowledge with new feedback.

The occurrence of local optimum points can be another challenge in reinforcement learning. Here, the agent executes the tasks to obtain a high reward, without executing the task in the correct way to achieve the target. For example, consider a game of a race. The model can take actions to gain rewards without ending the race.

5.3   FEDERATED LEARNING

5.3.1   OVERVIEW OF FEDERATED LEARNING

Generally, machine learning-based applications process a large amount of data, and it is important to avoid data breaches. It keeps data in one central location. Therefore, privacy preservation should be incorporated into such applications. Federated learning (FL) introduces training in machine learning by transmitting model replicas to the locations that perform data training. In other words, the learning algorithms are trained over many distributed edge devices by keeping a local data instance without replacing them. Therefore, this will eliminate the requirement of moving big data records to a central device for training. Therefore, federated learning, also known as collaborative learning, supports distributed and heterogeneous networks with pre-serving data privacy.

Figure 5.9 shows an overview of federated learning with other learning approaches, where federated learning is a deviation from distributed learning. Centralized learning

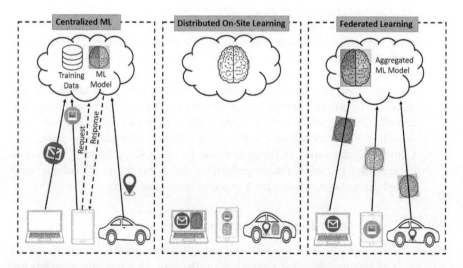

FIGURE 5.9   Centralized vs distributed on-site vs federated learning.

passes the data to a central location, such as a cloud, to train the model. Clients use APIs to access the model through services. In distributed on-site learning, a model with a local dataset is created in each device. Initially, a model is distributed to each of the devices from the central location. After that, the devices can perform standalone, without communicating with the central cloud server. As an extension, federated learning trains the model in each edge device and passes its parameters to the central location to aggregate. Here, data is stored on edge devices and only knowledge sharing occurs among the aggregated models.

Accordingly, data is kept in a central location and the model training is distributed to edge devices in distributed learning. Whereas federated learning trains a part of the model by keeping a portion of data in local devices and passing the parameters between the aggregated devices allowing collaborative learning. It trains the models across distributed datasets and preserves the sensitive information in local devices, as data is not communicated via the network. This helps to reduce the cost associated with data transferring as well.

5.3.2 Federated Learning Process

Supporting device-level training, federated learning is based on splitting the iterative learning into a set of interactions between the central location and different edge devices, as shown in Figure 5.10. Here, the edge devices carry out the local training based on the guidance of the central location. During each iteration, the present state of the global model is communicated to the associated edge devices. These local nodes then train the local models and the generated model updates in each of these edge devices are sent for combination into a single update on the global model. Thus, the central location performs the aggregation of the model based on the updates of the local models.

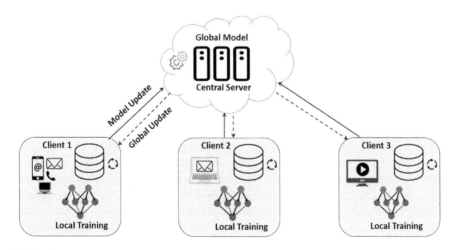

FIGURE 5.10   Federated learning architecture.

In detail, the data remains at source devices, which are known as clients, and they get a replica of the global model from a central location. Then this global model will be trained at each device using the local data. Here, instead of keeping a single global dataset, federated learning distributes many model versions among devices with local data, which will train locally.

Therefore, federated learning trains a set of models in multiple client devices and aggregates the knowledge generated from each model, which is sent to a single final model at the central location. This information is sent using parameter sharing through an encrypted communication channel. After that, the weights of each local model at the client are updated in each epoch, to continue with the model training in edge devices and pass the knowledge again to the server or cloud. The server aggregates the model updates to improve the combined model preserving the privacy of the data and this process repeats. The final model will behave as it was trained using a single dataset. Thus, the key advantage is the central location does not keep individual updates and the data will remain on the client's devices.

The key steps in federated learning can be listed as follows. This can be an iteration with initialization, client selection, configuration, reporting, and termination.

Step 1: select the underline model framework that supports FL. (Initialization)

The selection rules to implement a model are based on the aspects such as the data type, compliance of the proposed framework such as TensorFlow and PyTorch with the infrastructure and the feasibility of applying a given technology.

Step 2: determine the network mechanism. (Initialization)

This considers the format of the communication and the framework for sending the rules among the local devices. Few options would be PySyft with PyTorch for lower-level access to modeling operations, Flower, which supports multiple modeling frameworks, and Tensorflow Federated. At this stage, the local devices are initialized, activated and stay until they get the tasks to do from the central location.

Step 3: establish a central location to manage client services. (Client selection)

It is necessary to coordinate the communication between the participants and monitor the training progress to ensure maintainability, reliability, flexibility, and reliability. Additionally, the service operations consider aspects, such as authentication and authorization mechanisms. This considers having a stateless service for load balancing, such that deciding a storage method to keep the transitional knowledge among the edge devices. The service functionalities consider aspects, such as authorization or service isolation between different data networks, clients' interaction with the training, test statistics, and quality metrics.

Step 4: design the client system. (Client selection)

The client system should carry out local model training and share the knowledge-based parameters with the central location through services. The parameter sharing helps to update the models in local machines. The training process may start with a selected set of local devices while the rest of the devices stay until the next round of federated learning. The associated considerations are, the type of package (installable or docker image), managing dependency versioning, client authentication, and server communication, observing model training and handling errors.

Step 5: design the training process. (Configuration)

It is important to identify the private data used by each device for local model training. Here, the central service manages the related meta-data, such as the availability of datasets by different clients and the datasets used by each client model training. In this stage, the central device assigns a set of devices to start training those results in the update of mini-batches.

Step 6: create a method to manage the model. (Reporting)

We need to manage the access rights and the measures associated with a model. It is used to select the users that can train the model. The key considerations are the access rights of each client and the model storage location. At this stage, when each device transfers the local knowledge to the server, it aggregates the model and communicates the updates back to the local devices. It also manages the failures due to disconnected devices and lost updates. It starts the next federated round by selecting a device set again.

Step 7: security management and privacy preservation. (Termination)

Since the model trains by parameter sharing of different edge devices, the final model is accessible for each client locally. It is important to identify the acceptable risks, such as the possibility of identifying the clients who performed a given part of model training. Such information sharing can be prevented by incorporating privacy-preserving mechanisms into the trained weights, before transmitting to the central location. However, the optimization of the risk associated with a model trade-off with the model performance. Subsequently, the central device aggregates the knowledge received by different local devices and finalizes the model when termination criteria, such as the completion of the iterations or achieving an accuracy beyond the threshold are met.

5.3.3 Types and Properties of Federated Learning

The main categories of federated learning are listed as follows.

- Centralized: different tasks of the learning algorithm and the edge devices are controlled by the central device. All the clients selected by the server, communicate the knowledge to the single server, thus the server may get into a bottleneck situation and tend to a single point of failure.
- Decentralized: the edge devices are organized among themselves such that the updates of the models are shared among the interconnected edge devices to generate the final model without a central server. Thus, it addresses the problem of single-point failure. However, based on the selected topology of the network (e.g., Blockchain based) can impact model learning.
- Heterogeneous: use a diverse set of client machines are used with different computation and communication capabilities such as mobile and IoT devices.

The following aspects impact the properties of federated learning.

- Data partitioning.
  Several data partitioning techniques are available in federated learning. In horizontal data partitioning, similar features with a slight intersection of the sample space are included in different local machines. This is a widely used method, where all the clients use a shared mode leading to an easy aggregation process at the central machine. Considering related examples of different partition types, a patient dataset for a given type of disease in a hospital is a possible application for horizontal data partitioning. In vertical data partition, different feature spaces with the same sample space are used by the clients. We use techniques, such as entity alignment, to identify the overlapping samples in the client data that are used for training. For example, student GPA datasets from universities in different countries can be considered, where the feature space would be the grading scale and evaluation metric. Hybrid data partitioning is a combination of the above two methods. A possible application would be measuring student performance across branches of a set of universities.
- Machine learning model.
  The selection of the machine learning model based on the dataset and the associated task. In federated learning, homogeneous models use the same model in all clients and the aggregation of gradients in the server. In heterogeneous models, each client has a different model, thus no aggregation method, but contains ensemble methods such as max voting.
- Privacy mechanism.
  To avoid information leakage amongst clients, the server interprets clients' data using learning gradients without encrypting data. Differential privacy methods such as adding random noise to the model parameter or associated data, are used to hide or mask the gradients. However, the noise can result in low model accuracy. Cryptographic techniques such as secure multi-party computation

and homomorphic encryption sent the encrypted data from local devices to the central device. The central server decrypts the encrypted output to get the final result. However, they are computationally expensive.

- Communication architecture.
  In federated learning architectures, the functioning is the same, but the client-server communication is different. In the centralized architecture, the central device updates the parameters shared by the local devices. In the decentralized architecture, a given local machine is randomly selected as a server for each epoch, to update the global model and communicate with other clients in the network. The implementation is complex and consists of peer-to-peer (P2P), graphs, and blockchains.
- Scale of federation.
  The scale of the federation falls into two categories. The cross-silo category has a small set of clients with large computational abilities. This can correspond to an organization and has high reliability as it is always available to train. The cross-device category has a large client count with a small computation power. This can be associated with mobile phones and has low reliability as the low network can hinder the availability of the device.

5.3.4 Applications of Federated Learning

At present, many software applications collect user data and users have concerns about sending their details to a central location for reasons such as privacy, and usage of the data bandwidth. With the increase of the device's computational power, these data are stored in local devices and trained data using federated learning. This enables the predictive features of the application by preserving user privacy. For example, Google and Apple use this approach to create learning models with distributed datasets preserving user privacy. Some of the applications that use federated learning are listed as follows.

- Learning over smartphones: statistical models that learn through user behavior to detect faces, recognize the voice, and predict the next word using mobile phones.
- Google's Android keyboard improves word recommendation without uploading data to the cloud and training the model using user interactions with mobile devices. G-Board can personalize the user experience by the individuals' way of using the phone by referring to device history and suggesting improvements.
- Learning among organizations: institutes such as hospitals operate with many patient data that should preserve privacy. These data are stored locally, due to the associated ethical, administrative, and legal constraints.
- Predict human body conditions such as strokes using wearable devices.
- Identify the behavior of pedestrians and other vehicles in self-driving vehicles.
- Robo automation consists of NLP models that use data from a different locations.

- Credit card fraud detection systems in financial organizations.
- Recommender and personalization systems that use data from different consumers.
- Computer-vision-based healthcare diagnostics systems in different hospitals.
- Enhancements in Apple's Face ID and Siri voice recognition.

5.3.5 Challenges of Federated Learning

Generally, the aspects, such as data privacy preservation, optimizing distributed processing, and handling large-scale model training, are some of the challenges associated with federated learning. Other than that, federated learning raises several challenges as listed below.

- Expensive communication: since federated networks comprise many devices, network communication can be slower and more expensive than local computation. Thus, communication-efficient methods need to be developed for the repeating transmission of small messages or model updates, instead of transmitting the entire dataset during model training.
- Systems heterogeneity: the aspects such as the hardware configuration, power and network strength of each of the local device is different. This results in having different storage capacities, communication and computational capabilities in each device. Based on these constraints, only a set of devices are active during one iteration. These settings are more error-prone and device failures can happen, which leads to loss of model updates during training.
- Statistical heterogeneity: data is created and acquired by different distributed devices. Consider the NLP task of predicting the new word among mobile phone users with different languages. Although different data points exist among devices, there is a method to identify the relationship among the devices and the distributions as the basis. This can result in issues, such as insufficient data labels in the client machines and processing complexities. The data heterogeneity can result in bias and different sizes in local data. Therefore, preserving data privacy is hard to achieve with the statistical discrepancy in the data and the configuration constraints in the local devices. Since the local data can change over time, it may result in temporal heterogeneity and may require interoperability, and regular curation.
- Privacy concerns: federated learning protects data by sharing the knowledge learned from the model and related updates, such as gradient, without sending the original data. This communication may result in identifying local information, such as location and IP address by the central device. Also, the performance of the model may decrease by using privacy-preserving methods, such as secure multi-party computation or differential privacy. Therefore, a balance is needed between privacy and performance aspects.

5.4   MULTI-MODELING WITH ENSEMBLE LEARNING

5.4.1   OVERVIEW OF ENSEMBLE LEARNING

Ensemble learning combines multiple machine learning algorithms into a reliable model to improve the performance of existing models. Although individual models perform well, in some scenarios the integration of a set of models produces better results using an ensemble approach for a given task. Overall, in ensemble learning the models are stacked together to perform multi-layer processing to obtain a robust model, as shown in Figure 5.11. It integrates the results of a set of models to decrease the prediction variance and the generalization error, thereby increasing the precision of learning and discovering a new architecture.

Let us distinguish ensemble learning from traditional deep learning. In general, the concept of fusion combines different embeddings to produce a single entity. The integration of data from different sources and generating specific results is known as data fusion. We learnt that neural networks are non-linear methods and CNN is an efficient model. They are flexible and scalable with the size of the dataset. However, since they train with stochastic learning algorithms, most of the time these models may highly depend on the considered data type resulting in high variance. Also, the weight assignments may change in each training process, which will result in varying predictions. Accordingly, when the process is based on one model and sensitive to the training data, it does not generate high prediction results.

This can be addressed by training the data using a set of models to decrease the variance and integrate the results. Those base learner models can follow different procedures of mapping the features with variant decision boundaries. Thus ensemble learning can produce better results. This improvement occurs by combining multiple features with heterogeneous large datasets, which are extracted from multiple classifiers.

5.4.2   ENSEMBLE LEARNING PROCESS

Ensemble learning uses a set of models to train a given dataset and combines the results of each model to generate the final prediction. Let us consider an example as shown in Figure 5.12. Here, the four main layers of a CNN, namely the convolution, activation, pooling, and fully connected layer, work together to provide a prediction.

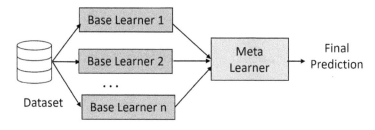

FIGURE 5.11   Overview of ensemble leering.

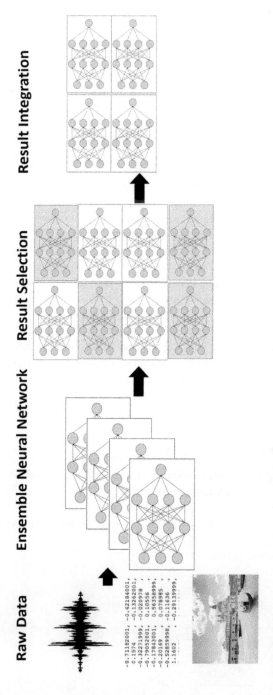

FIGURE 5.12 Process overview of ensemble learning.

Three-dimensional images are input to the model and the CNN groups the pixels and processes using filters. Based on the complexity of the dataset, we can decide the number of filters to be used. The pooling layer is used to decrease the parameter space of the input through regression. This process is applied repeatedly on a given dataset to produce a reliable result.

Generally, we use a small set of models such as three, five, or ten trained models, to avoid unnecessary computational expense and decline in performance. The weight initialization of the ensemble model is proportionate to the prediction accuracy of the individual models. These weights are assigned in a way that reduces the MSE of the total of weighted models, for each iteration. Equation (5.1) states the production of the ensemble model, where w_i and y_i denote the calculated weight and result of model i, respectively. Then the weights are tuned to obtain the minimum MSE for the ensemble model $(w_1y_1 + w_2y_2 + \cdots + w_jy_j)$, which denotes the addition of bias and the variance of the models.

$$EM = \sum_{i=1}^{n} \left(w_i . y_i \right) \tag{5.1}$$

The bias is defined as the deviation between the expected prediction and the actual values as stated in (5.2). The variance and the total expected prediction error (MSE) are defined in (5.3) and (5.4), respectively.

$$\text{Bias}[m'(x)] = E[m'(x)] - m(x) \tag{5.2}$$

$$\text{Var}[m'(x)] = E[m'(x)^2] - E[m'(x)]^2 \tag{5.3}$$

$$E[m(x) - m'(x)^2] = (\text{Bias}[m'(x)])^2 + \text{Var}[m'(x)] = \text{Var}(\epsilon) \tag{5.4}$$

where
$m'(x)$: output of a base model $m(x)$
$E[m'(x)]$: expected output error of model m
$m(x)$: actual class values
$\text{Var}(\epsilon)$: obvious error based on the noise variance

The ensemble learning process can be designed by changing the following elements.

- Changing training data

We can use mechanisms to alter the training data in each model in ensemble learning. As a simple technique, k-fold cross-validation can be used to measure the generalization error. Here, the training dataset is divided into k subsets and each subset is trained using a unique model. Bootstrap aggregation (bagging) is another technique, which trains the model with a resampled training dataset with a replacement, which results in different generalization errors. Random training subset ensemble is another technique.

- Changing models

The configuration of each model used for the ensemble can be varied. For instance, models differ based on the number of layers or nodes, learning rates or regularization types. Thus, the ensemble model learns a heterogeneous mapping function and results in a smaller correlation in the output.

- Changing combinations

The combination method of the outcomes from ensemble models can be varied. As a simple technique, model blending can be used that calculates the average of the predictions of each model. This weighted average ensemble can be improved by taking the optimized weights techniques, such as hold-out validation. Additionally, new models can be learned using techniques such as stacking. As an extended technique, boosting can be used. It incorporates one model at a time for the ensemble learning process to address the errors that occurred during the previous model training. Model weight averaging is another method that combines the weights of several models that has a similar structure.

5.4.3   ENSEMBLE LEARNING TECHNIQUES

We have learnt that the objective of ensemble learning is to decrease the variance of the predicted results and the generalization error by merging the outputs of several learning models. As discussed above, different ensemble learning techniques are obtained by varying training data, and models, and combining the mechanism of the predictions. The selection of the ensemble techniques depends on the application. There are advanced ensemble learning techniques such as bagging, boosting, and stacking as described as follows.

- Bagging (bootstrap aggregation)

The bagging method selects individual data points more than once and performs a random sample of training data, to reduce the prediction variance within a noisy dataset. Here, a replacement technique is used to create random subsets of a dataset, which are used as independent datasets that are used to train models in parallel. Therefore, a given data point can belong to many subsets of data. The testing phase considers the output of the models that are trained using different subsets of a given dataset, as shown in Figure 5.13. The final result is obtained by passing the several model outputs through an aggregation process. Bagging shows higher bias, such that the predictors are less correlated and reduce the ensemble's variance.

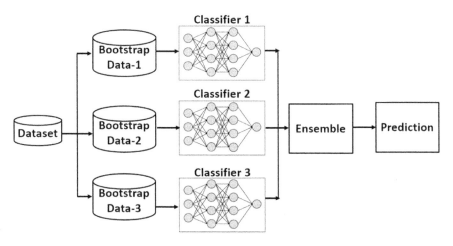

FIGURE 5.13 Process of bagging technique with parallel processing.

- Boosting

Generally, each model would not perform well on the entire dataset; however, they do have high performances in parts of the dataset. Boosting processes, the dataset sequentially, such that the entire dataset is input to the initial model and analyze the result. The data points that are incorrectly classified by the model are then passed to the second model. Thus, the second sub-model focuses on the challenging regions of feature space and learns an applicable decision boundary. As the name suggests, each model will contribute to boosting the performance of the overall ensemble. Subsequently, the same process is followed and the combination of all the previous models is used to generate the final result on unseen data as shown in Figure 5.14. Here, each model is dependent on the previous model as the successive models aim to rectify the errors of the prior model training for a given subset of data. Thus, the boosting algorithm ensembles a set of inadequate models to generate a robust learning model and boosts the entire prediction using the majority voting weight. This method reduces the bias in the ensemble prediction. Thus, the selected classifiers should be simple models with less trainable parameters that result in low variance and high bias.

- Stacking

The stacking technique uses bootstrapped data subsets and trains multiple models parallelly as in the bagging technique. The output of all these models is used to combine the multiple models through a meta-classifier that produces the overall prediction. Here, two layers of classifiers are used to ensure proper training. The meta-classifier in the next layer may capture the missing features from the set of models in the first layer. For instance, the output class assignment probabilities produced by the first layer of models can be averaged with some weights to combine the model outputs. Then the argmax over the average predicted class probabilities can be used for the

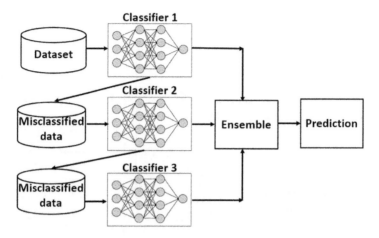

FIGURE 5.14   Process of boosting technique with sequence processing.

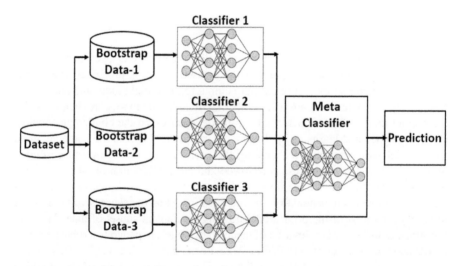

FIGURE 5.15   Process of stacking technique with meta-classifier.

final prediction. A stacking with one level is shown in Figure 5.15. Additionally, there are ensemble models with multi-level stacking, where extra classification layers are included among them. However, compared to the relatively low improvement of the performance results, such methods consume more computational costs.

- Mixture of experts

This technique uses several classifier models, and their outputs are ensembled using a generalized linear rule. A gating network, which is a trainable neural network, is used to decide the weight assignment for the combinations as shown in Figure 5.16.

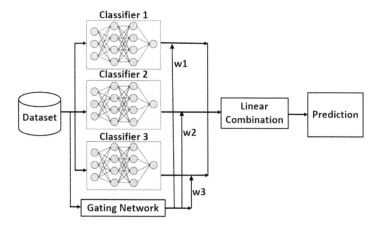

FIGURE 5.16   Process of the mixture of expert techniques with a generalized linear rule.

The mixture of experts' method is applied when there are various models trained on different classes of the feature space, supporting the information fusion problem.

The following are basic techniques in ensemble learning that operate on similar data.

- Max rule: this technique is based on the probability distributions produced by each model. The prediction of the ensemble model is the class label that has the maximum confidence score of the output among the classifiers, so that it can be applied for multi-class classification as well.
- Majority voting: this method considers a random set of classifier models and computes the predictions for each sample. The most predicted class label by many models is considered as the output of ensemble learning. Since the majority-voted class is selected, this technique works well for binary classification. In multi-class classification, if two or more classes have the same highest voting, then a random class is selected as the final prediction.
- Probability averaging: this method first computes the probability scores, which is the confidence in prediction, for all the models separately. Then it calculates the average scores of all the involved models, considering all the class labels in a given dataset. Among the average scores, the class label with the maximum probability is considered as the predicted class of the ensemble model.
- Weighted probability averaging: this method considers the weighted average of the probability (confidence scores) from multiple classifiers. In the classification process, some classier models perform well, and some are not. The weights are based on the importance of each classifier, such that a high weight is assigned to the classifier with high performance. Thus, the result of ensemble learning gets better predictive ability.

5.4.4 Applications of Ensemble Learning

Let us discuss some of the scenarios that can be used in ensemble learning.

- Unavailability of a single optimal model: a model may well adapt to predict only some classes within the given dataset. A different model may perform better on another set of classes within the same dataset. An ensemble model provides a more sensible decision boundary for the classification classes.
- Extra or insufficient data: when there are more data, different models can be used to classify sections of data, and ensemble them during prediction time. This may reduce the cost and computation power of training the entire large dataset using one classifier.
- Similarly, when there is a lack of data, techniques such as bootstrapping can be used for ensemble learning. Here, a given dataset is split into a set of subsets utilizing replacement methods, such that a given data instance may belong to many subsets.
- Requirement of confidence estimation: in some scenarios, it is required to consider more important models with a high prediction confidence. The usage of the confidence scores of the individual classifiers in the ensemble model provides better results compared to the predictions of the majority set of models.
- High problem complexity: some applications contain complicated decision boundaries, where a single model may fail to predict accurate results.
- Information fusion for classification performance improvements: Different distributions of the subsets of data that belong to the same set of class labels are trained to generate robust decisions.

Following are some of the applications of ensemble learning.

- Disease detection: lung disease detection using chest X-ray and CT scans.
- Social networking: use face detection and recognition to tag users.
- Legal, banking, insurance: provide optical character recognition.
- Remote sensing: different sensor devices produce a variety of data consisting of different resolutions.
- Entertainment: filter functionalities in social media networks.
- Document digitization: enables flexible access to documents.
- Landslide detection.
- Scene classification.
- Land cover detection.
- Credit card fraud detection.
- Speech emotion recognition in multi-lingual environments.

REVIEW QUESTIONS

1. Compare and contrast transfer learning and fine-tuning.
2. When should deep transfer learning be used?

3. What is the purpose of using reinforcement learning?
4. What are the challenges in reinforcement learning?
5. Why is federated learning important?
6. What are the considerations to select an ensemble learning technique?

6 Enhancement of Deep Learning Architectures

6.1 MODEL PERFORMANCE IMPROVEMENT

Deep learning algorithms are based on both predictive modeling and statistics. Enhancing the performance of a model can be challenging and mainly depend on the data type and the model training with optimal hyperparameter tuning. Several methods are utilized to create predictive models. Although there are not specific rules, both theoretical and practical experience assist to produce a model with better performance. However, sometimes there are models with higher training accuracy but low testing data accuracy, which result in overfitting. Let us identify possible strategies to improve model performances.

- Train with more data

Deep learning models mainly depend on data. The validation accuracy of the model can be improved by adding more data. For instance, in image datasets, the diversity of the available dataset can be increased using data augmentation. General methods include image flipping along the axis and introducing noise. In addition, advanced methods such as generative adversarial networks (GANs) support data augmentation.

- Treat outliers and missing data

The real-world data often contain some absent data and outliers. This leads to a decrease in the performance of the model by introducing model biases. Thus, it should be handled during the preprocessing of the dataset. For instance, when the dataset contains continuous data, we can use statistical measures, such as mean, median, and mode as a substitution for missing data. When the dataset contains categorical data, the values are considered as a separate class. Additionally, models can be built to predict the missing values. The outliers can be handled by applying methods, such as deleting the data, performing alterations, binning, or treating outliers separately.

DOI: 10.1201/9781003390824-6

- Feature engineering

Feature engineering extracts more information as new features from the available data. The identified features are used to explain the data and to improve prediction accuracy. In this process, feature creation derives new variables from available data to identify the connections of data points, then it transforms the features to the next process. Feature transformation is supported by different techniques, such as normalization, which changes the scale of the data. For instance, data can transform to a range between zero and one to get the variable in the same scale. The skewness of variables in normally distributed data can be removed using methods, such as square root, log, or inverse of the data. Also, data discretization is used to transform the numeric data into discrete by dividing data into bins.

- Feature selection

Feature selection finds a subset of features that can describe the connection between the source and target data in an effective way. This is supported by the knowledge of the considered domain and experience, visualization of the relationship between variables, and statistical parameters (e.g., p-values). Additionally, there are dimensionality reduction methods such as principal component analysis (PAC) that represent training data in a reduced dimensional space, but still, characterize the intrinsic connections in the data. Other methods can be considered as maximum correlation, low variance, factor analysis, and backward/forward feature selection.

- Add more layers

The addition of more layers to a neural network improves feature learning capabilities. It allows us to learn the features of the dataset more deeply by recognizing subtle differences in the data. However, it depends on the nature of the application domain. For instance, if the data classes have a clear difference, then it is sufficient to have a few layers. However, if the classes are slightly different with fine-grade features, then more layers are needed to learn subtle features that differentiate the classes.

- Algorithm tuning

The hyperparameter tuning majorly influences the outcome of the learning process by finding the best value for each parameter. Knowledge of the parameter type and its impact is useful to make decisions to improve the model performance.

- Change image size

Feasible image size should be identified during the data preprocessing. When the image is very small in size, it is hard to learn distinctive features that are needed to recognize the image. In contrast, if the image is too large, the model requires more computational resources to process, otherwise, the model may not be sufficient to learn the data. Converting an image from a small resolution to a large size image results in pixelation, which is the visibility of individual square pixels that make up an image. This leads the image to have blurry sections or fuzziness, which causes negative effects on the model performance.

- Increase epochs

The number of iterations the entire dataset goes through in the model is defined as an epoch. When we increase the number of epochs, the model will train incrementally. Generally, epochs are increased when there is a sufficiently large dataset. Subsequently, when the model no longer increases the accuracy while increasing the epochs, we consider the model learning rate at this point. This hyperparameter decides whether the model reaches its global minimum or stays at a local minimum.

- Decrease the number of color channels

Generally, color channels are used to represent the image dimensionality. For instance, color images in RGB contain three color channels. In contrast, images with grayscale have one channel. When the color channel is dense, the dataset becomes complex too and will take more time for model training. Based on the application domain, if the color does not impact the prediction, then it can be converted to grayscale and processed, which requires fewer resources.

- Transfer learning

Transfer learning uses pretrained models that were trained on large datasets for the predictions of a new task. We have discussed this topic in Chapter 5.

- Ensemble methods

Ensemble learning is a widely used approach to combine the output of a set of poorly performing models and leads to better predictions. Different techniques, such as bagging (bootstrap aggregating), boosting, and stacking, support to increase model performance. Generally, ensemble models are more complex than conventional models, which is the basis for ensemble learning.

- Evolutionary algorithms

Evolutionary or genetic algorithms aim to replicate the evolution of the population in nature to optimize functions. Given an initial set of predictions, first, it evaluates the appropriateness of each prediction and selects the population fittest to create offspring. Then it creates offspring and mutations by mixing and matching the genes in different variables of each solution. The mutation is a local operation that changes the behavior. For instance, a change of a hyperparameter or the addition of a new layer is considered as mutation. Once the model training happens, the process assesses the prediction, then feeds the output back to the input, and repeats the process until the expected output is generated.

- One-shot model or supernet

This is an individual large model consisting of all the possible operations. It produces weights that can be used by other potential models. After the training process, it samples sub-architectures and compares them with the validation data. It uses the advantage of parameter sharing to its maximum. Generally, this model trains with gradient descent by transforming the space into a continuous and differentiable form.

- Cross-validation

Cross-validation helps to achieve a more generalized relationship in data modeling. It leaves a data sample without training the model and performs testing on that sample of data.

6.2   REGULARIZATION

Computational models are known for their hard interpretability, which often identifies them as black boxes where the input data is fed and then an output is generated from the model by executing its classification or regression algorithm. The model training can be a complex learning process and it can be challenging to generalize the model, depending on the input data. This phenomenon is known as overfitting, which leads to low accuracy. Here, the model is adjusted to the peculiarities of the dataset where it cannot generalize well into new data. This becomes more complex when the model tries to detect noise in the input data, which reflects the data points that do not indicate the real features of the data but arbitrary chances. Figure 6.1 shows different model-fitting scenarios, which we discussed in earlier chapters. This can be addressed

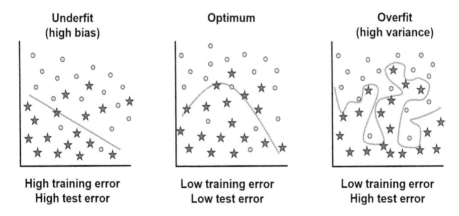

FIGURE 6.1   Model fitting scenarios.

by applying cross-validation that estimates the error over the test set and helps to generalize by deciding the parameters that would work best for the model.

Regularization is an important concept in learning algorithms to reduce loss and avoid overfitting or underfitting. This approach adds extra information to data and helps to fit the model on an unseen dataset with reduced errors. This is a type of regression that decreases the coefficient closer to zero. Because of that, regularization does not support learning complex models and reduces overfitting. This simplifies the model by using a smaller number of parameters. Since there are no regression coefficients with higher values, it prevents overfitting. This method minimizes the loss and complexity together resulting in a more streamlined and parsimonious model that performs better at predictions.

The model complexity can be represented as the following functions.

- Weights of all the features
- Total of features with non-zero weights.

The regularization techniques can be explained as follows.

- L1 regularization

L1 regularization (or lasso regression) focus on model simplicity by subsequently making the weights of the unimportant features zero, that result in discarding some features. This is done by considering the total of the absolute values of the weights and reducing a slight value from the weights in each iteration. In other words, the regularization term is multiplied by a small value and adds the entire value to the loss function as a penalty. When there are many features it results in a sparse solution, and it is computationally advantageous to ignore the features with zero coefficients. Accordingly, it tries to assign more weight to zero to increase the sparsity in the weights, rather than lowering the average of the total weights, as in the case of L2 regularization.

- L2 regularization

L2 regularization (or ridge regression) is another method to address overfitting by introducing a penalty to the loss function. This is mainly used for models that have collinear or mutually dependent features. This is a commonly used method as it performs well enough to alleviate the overfitting problem compared to L1 regularization. In L2 regularization, weights near zero have a low impact and the outlier weights have a high impact on the model complexity. This process calculates the total of the squared weights of all the features and sums it up with the square difference between the expected and actual results. This reduces the weights to lower values avoiding the complex learning of the model for a given feature and this addresses the overfitting problem. This regularizer affects the weights more with less influence on the loss function. Therefore, the variance of a model can be reduced to generalize on unseen data.

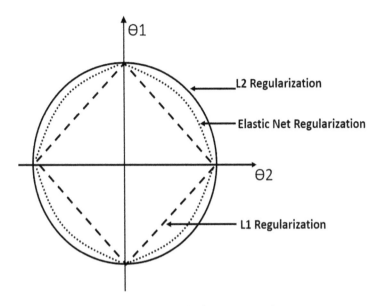

FIGURE 6.2   L1, L2, and elastic net regularization representation.

- Elastic nets

L1 regularization's characteristic of making the weights closer to zero helps efficient learning. However, since the technique discards unimportant features, which can lead to removing some connections between features, the model tends to be less generalizable. In contrast, L2 tries to retain the features by preserving the connections between the features. As a result, although L2 is more generalizable, it becomes a dense network. Considering the advantages of L1 and L2 regularization techniques, the elastic net methods combine both, as shown in Figure 6.2.

- Early stopping

Early stopping is used in model learning with iterative approaches like gradient descent. Since these techniques can reduce the model generalizability, the early stopping mechanism states the maximum number of iterations that the training should be executed before the model gets overfit. This method initially sets many epochs to train and when the performance of the learning curves starts to reduce, it stops training on the validation data. The model tries to follow the loss function exhaustively on the training data, by tuning the parameters. Accordingly, by monitoring the loss function on the validation data, if there is no improvement in the validation set, then the learning stops without processing all the epochs. Figure 6.3 shows a representation of early stopping. The early stopping regularization has been useful in providing consistency in boosting algorithms, such as Adaboost.

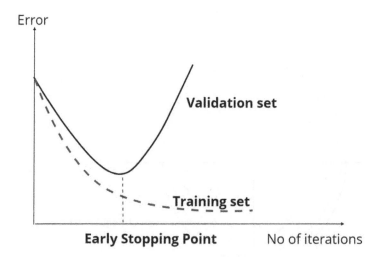

FIGURE 6.3   Early stopping representation.

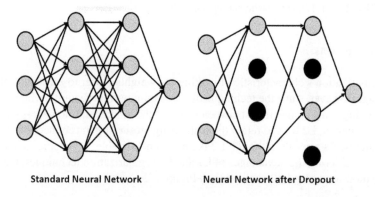

FIGURE 6.4   Neural network architecture before and after dropout is performed.

- Dropout

The dropout technique randomly ignores or drop-out some number of layer outputs at different points during training. Subsequently, this is considered as a layer with a diverse set of nodes and connections to the previous layer, supporting model training with different architectures simultaneously. Thus, each time when the gradient is updated, a new sparse model is generated without the dropped nodes based on hyperparameters. This change may help network layers to correct mistakes from prior layers and produce a robust model. Dropout can cause the training process to become noisy based on the impact of the nodes for the inputs. Figure 6.4 shows an instance of a dropout network.

The models with dropouts may have a lower classification error with test data, because of their simplistic representation and generalizability. Since overfitting can be addressed using the dropout technique, which reduces the dependability of the nodes among different hidden layers, it results in low error values. However, this technique can take more training time. This can be addressed by using a regularizer, which is similar to a dropout layer. For example, dropout can be considered as an extension of L2 regularization for linear regression models.

6.3 AUGMENTATION

Generally, it is challenging to obtain a perfect dataset with the correct amount of data for machine learning model training. Although many public datasets support object detection and image classification, it is difficult for acute sufficient data to train models for a specific application domain. Training with insufficient data can cause model overfitting and underfitting. Thereby giving less accuracy in the test dataset because of the limited variety of the data, which enables the model to learn the hidden peculiarities of the desired application.

Data augmentation is a widely applied technique that expands the variety of the training data by utilizing realistic and random transformations. With minor alterations to the existing dataset, the model identifies those machine-generated images as distinct images, which will increase the variation of the training. This supports the model to identify the peculiarities of the dataset and performs well with the unseen data.

The following strategies support data augmentation.

- Basic data manipulations: this applies geometric transformations to data in the input space. The techniques associated with this category can be named as a rescaling of an image to a given rescaling factor, random flipping (horizontal and vertical), random rotation, cropping, shearing, shifting (height shift and width shift), change brightness, change contrast, translations, image color modification, image mixing to blend two images to a single image.
- Expansion of feature space: this applies transformations to the feature space. For instance, the latent representation of the original data unit can obtain using an autoencoder and add noise to transform the raw data.
- Augmentation with generative adversarial networks (GAN): considers a natural selection to increase data, as it works extremely well on data synthesizing. Some of the widely used techniques are pix2pix and cycleGAN. Pix2Pix GAN is a conditional GAN (cGAN), which utilizes real data, noise, and labels to create images. CycleGAN solves the image-to-image translation by mapping the source to the target by utilizing the trained aligned image pairs. However, it is not always to get paired images.
- Meta-learning: neural network models can be used to optimize other models with hyperparameter tuning to expand their layout. Here, a classification network can be used for the enhancement of an augmentation model, such as GAN, that produces images. Then, both the raw image and the augmented image send to another model to compare and assess the augmented image.

FIGURE 6.5   Different augmentation representations.

In most cases, it is appropriate to use a combination of multiple techniques in a single data augmentation to produce better new images. However, the class of a dataset should be considered when applying different augmentation techniques. For instance, in the number recognition model, applying flipped over the y-axis for an image with the number 3 does not represent a meaningful data instance as shown in Figure 6.5.

6.4   NORMALIZATION

In deep learning, normalization techniques are applied to transform data in such a way that all the features in a dataset lie on a similar scale. Generally, we use a range between zero and one to normalize data. If the data is not normalized, it will cause a dilution in the effectiveness of important attributes that are on a lower scale. The normalization layers support efficient and steady model learning. Let us discuss some of the normalization techniques, such as batch normalization, used to train with large batches without recurrent connections, and techniques to train with small batches such as weight normalization, weight standardization, group normalization, and layer normalization.

• Batch normalization

Batch normalization standardizes the inputs to a given layer. Generally, the neural network learning process tries to decrease the weight parameters in the direction given by the gradient, which depends on the current inputs. As the network layers are stacked together, a small weight update of the previous layer results in a large change in the connections from the input layer to other layers. This causes sub-optimal outputs to be generated from the current gradient. Since batch normalization limits the inputs of a given layer, such as the activations from the prior layer, the weights can be updated with better gradients and result in steady and efficient training.

The batch normalization layer performs the transformation by subtracting the input mean from the input of the current mini-batch and dividing by the standard deviation. Normally, most of the layer inputs have nearly zero mean and unit variance. Depending on the task the model may perform well with different values of means and variance. Algorithm (6.1) states the batch normalization process of transforming the input x into the output y, using two learnable parameters γ and β.

Algorithm (6.1)
Input = values of x over a mini-batch: $\beta = \{ x_{1.....m} \}$;
     Parameter to be learned: γ, β

Output = $\{ y_i = BN_{\gamma, \beta}(x_i) \}$

$$\mu_\beta \leftarrow \frac{1}{m} \sum_{i=1}^{m} x_i \quad \text{//mini-batch mean}$$

$$\sigma_\beta^2 \leftarrow \frac{1}{m} \sum_{i=1}^{m} (x_i - \mu\beta)^2 \quad \text{// mini-batch variance}$$

$$\hat{x}_i \leftarrow \frac{x_i - \mu\beta}{\sqrt{\sigma_\beta^2 + \epsilon}} \quad \text{// normalize}$$

$$y_i \leftarrow \gamma \hat{x}_i + \beta \equiv BN_{\gamma, \beta}(x_i) \quad \text{// scale and shift}$$

Generally, the normalization layers are placed between a non-linear activation layer such as ReLU and other linear or convolutional or recurrent layers. In an activation layer since the activations are centered around zero. This supports training as it prevents non-active neurons occurred due to incorrect random initialization.

The limitations of batch normalization are as follows.

→ Require large batch size: since each training iteration of batch normalization determines the batch statistics, and the mean and variance of the mini-batch, a large batch size is needed to approximate the mean and variance from the mini-batch.

→ Does not perform well with RNNs due to complexity. Since there are recurrent connections to preceding timestamps, it uses distinct learnable parameters for each timestamp in the batch normalization layer.

→ Different training and test calculations. This increases the complexity as the batch normalization layer calculates a fixed mean and variance calculated from the training dataset, without calculating the mean and variance of the mini-batch from the test dataset.

• Weight normalization

Weight normalization decouples the length from the direction of the weight vector and reparameterizes to increase the training efficiency. This can use two parameters, the length and the direction of the weight vector, without taking the gradient descent. Weight normalization works with RNNs; however, this technique is significantly less stable and not commonly applied.

- Layer normalization

This technique considers the direction of the features to normalize the activations to zero mean and unit variance. Since it does not consider the direction of the mini-batch for the normalization, it addresses the limitations of batch normalization by ignoring the reliance on the batches. Thus, it can be applied to recurrent networks too.

- Group normalization

The features are split into groups and each group is normalized individually along the direction of features. This works better than layer normalization, as the hyperparameters are tuned based on the groups.

- Weight standardization

This technique transforms the weights that result in zero mean and unit variance. This is applicable for layers, such as convolution, linear, and recurrent. During the forward pass, it transforms the weights and calculates the corresponding activations. Generally, group normalization with weight standardization performs well even with small batch sizes. This is applicable for applications without large batch sizes as it consumes memory for dense prediction tasks.

Additionally, the following normalization techniques are used in practice.

- Z normalization: Z-score is a scaling type that considers the number of standard deviations located away from the mean of the distribution. This is useful when the data distribution does not contain extreme outliers.
- Min-max normalization: this will linearly rescale numerical feature values into zero mean and unit variance, which is the standard interval. It shifts the feature values so that their minimum value is 0 and divides them by the difference between the original maximum and minimum value (new maximum value). This scaling method is useful when the approximate values are available for the maximum and minimum data bounds with a smaller number of outliers. Also, the data points should be nearly uniformly distributed over the interval.
- Decimal scaling normalization: the feature values will be scaled in terms of decimal values. Thus, the values of attributes will be multiplied or divided with a power of 10.
- Log scaling: to transform a wide interval to a tight interval, this takes the log of the attribute. This method is useful when many data points consist of a small number of values, while few data points contain the remaining.
- Feature clipping: this is used when the dataset contains extreme outliers. This will cap all the attribute values that are above or lower to a certain value to a fixed value. Feature clipping can be applied after applying other normalization techniques.

6.5 HYPERPARAMETER TUNING

Hyperparameters are a set of parameters used to configure the machine learning model with the focus of lowering the cost function. This includes the number of nodes, learning rates, epochs, activation function, batch size, and optimizer. These can be defined as the variables that decide the structure and the training process. Initially, we set the hyperparameters before the weight-updating process in training. Hyperparameter tuning finds a set of the best possible values of these parameters during training. An optimal set of hyperparameter values result in high performance and low errors. Different datasets and machine learning algorithms require different sets of hyperparameters for accurate predictions. First, the hyperparameters are tuned in a model, and then the number of layers is tuned. The number of layers impacts the model accuracy such that fewer layers may result in data underfitting and more layers may lead to overfitting.

Following are the main hyperparameter types.

- Number of hidden layers: the increase in the number of layers can increase the accuracy. However, there is a trade-off between the simplicity of the model that leads to being fast and generalized and the accuracy of the prediction.
- Number of nodes in each layer: the increase in the number of neurons is useful up to a certain location. However, too wide layers may highly depend on the training data and may generate low accuracy predictions on unseen data causing overfitting. This should be decided based on the complexity of the problem.
- Learning rate: this controls the adjustment of the parameter size in each iteration. When the learning rate is high the model learns efficiently. However, it may result in local minima instead of reaching global minima. A low learning rate reduces the changes to parameter estimates and directs towards global minima. However, it requires a large number of epochs, with more time and computational resources.
- Momentum: by changing the values of the parameters frequently in their changing direction, this prevents local minima points and zig-zag movements in each iteration. Usually, it starts with low momentum and adjusts upward.
- An activation function: this affects the processing of the inputs of each layer to its corresponding output. The values that move between layers are changed based on the activation function used.
- Optimizer: the optimizer helps to change the weights and learning rate to reduce loss and achieve high accuracy.
- Batch size: this defines the number of subsets of the training set. When the training dataset is large, a batch size can be assigned without using the entire data at once. When the batch size is small, it learns fast. However, it results in a high variance in the unseen data.
- Epoch: this is the number of iterations the entire dataset goes through in the learning process. For instance, during a single epoch, the training set routes forward and backwards through the model once. Underfitting can occur, when we use a small number of epochs, due to insufficient learning. In contrast, many epochs can cause overfitting.

- Layer tuning: generally complex task solving requires many layers including regularization layers such as batch normalization and dropout to avoid overfitting. Generally, after the first few hidden layers, batch normalization is included that normalizes the input of each batch. The dropout rate defines the percentage of neurons to drop.

A search space must be defined before the algorithm, set bounds for all the hyperparameters, and add prior knowledge on them including setting a non-uniform distribution for the search. Different hyperparameter tuning algorithms categories are available as follows.

- Grid search

This technique automatically attempts all the different combinations of hyperparameters values and trains models for each and picks the model with the best performance. Looping through different values of hyperparameters and evaluating each combination is not computationally friendly as it increases the time complexity of the computation to optimize. In practice, the Scikit-Learn library in GridSearchCV implementation can be used. When the training model and list of hyperparameters are fed in a dictionary format, the function returns the performing model and its scoring metric.

- Randomized search

Randomized search trains model on random hyperparameter combination. The total number of combinations on which the several models are trained is less for randomized search compared to grid search. In practice, the Scikit-Learn package in RandomSearchCV implementation can be used.

- Halving grid search

Halving grid search is an optimized form of grid search hyperparameter. It searches over a specified list of hyperparameters using a successive halving approach. Initially, it assesses all the instances on a subset of data and repeatedly identifies the more suitable instances by processing larger subsets of data. This is less computationally expensive than the grid search approach. In practice, the Scikit-Learn library in HalvingGridSearch implementation can be used.

- Halving randomized search

Halving randomized search uses the same successive halving approach, and it is further optimized compared to halving grid search. Unlike halving grid search, it does not train on all combinations of hyperparameters instead it randomly picks a set of combinations of hyperparameters. In practice, the Scikit-Learn library in HalvingRandomizedSeachCV implementation can be used.

- Hyperopt-Sklearn

This is an extension of the Hyperopt library, which is an open-source Python library for Bayesian optimization and applies to models with many parameters for large-scale optimizations and over CPUs with many cores. Hyperopt-Sklearn supports an automated hyperparameter searching process of learning models.

- Bayes grid search

This technique utilizes Bayesian optimization to find the best parameters across the search space efficiently. It considers the structure of search space and samples the new instances based on the previous evaluations focusing on better predictions.

6.6 MODEL OPTIMIZATION

6.6.1 Overview of Model Optimization

The optimization process trains the model iteratively by adjusting the hyperparameters in every iteration until optimum results occurred. The optimization algorithms alter the model parameters like learning rate and weights, which are updated in each epoch or iteration, to reduce the loss losses (error function) or to maximize the efficiency of production. Usually, there is a trade-off between the network size and the speed, so when the number of layers increases it is difficult to apply optimizers. Different techniques like gradient descent (GD) and stochastic gradient descent (SGD) are used as optimization algorithms.

Let us revisit the following concepts.

- The maximum and minimum values of a function are denoted by maxima, and minima, respectively. Figure 6.6 shows the global maxima, local maxima, local minima, and global minima locations for a given curve. The global points are obtained for the entire domain of the function and the local points are based on a given range of the function. In a learning model, there are points for single global minimum and global maxima. However, there can be many local minima and local maxima locations.
- The learning rate denotes the step size at each iteration and is a tuneable parameter. Figure 6.2 shows the behavior of the learning rates. Since the trade-off between bias and variance is affected by the hyperparameters, a small learning rate can overfit the model with a large variance. In contrast, the model can underfit and result in a large bias, when the learning rate is high. The optimal values with a minimum loss can be identified using cross-validation.
- Gradients measure the slope of a function that considers the weight updates based on the loss. In general, the slope shifts its sign from positive to negative at minima. As a minimization algorithm, when the slope of points moved towards minima, the slope reduces as shown in Figure 6.3. When the gradient is increasing the slope got steeper and the model learns efficiently.

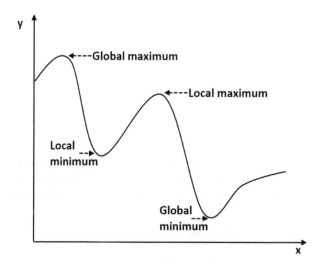

FIGURE 6.6 Minima and maxima locations.

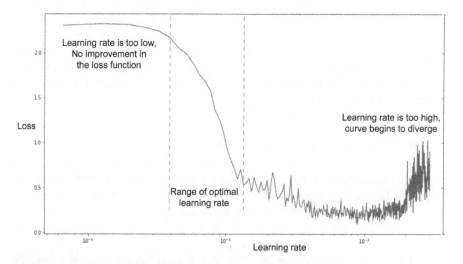

FIGURE 6.7 Representation of the learning rates.

- Exponential moving average (EMA)

The weights in a neural network need to update with a value. The consideration of the current gradient only may not give high-performance results. This can be addressed by aggregating the averages of all the current gradients and past gradients; however, it leads to having equally weighted gradients. In order to address this issue, an exponential moving average of gradients is considered. In this process, higher weight values are assigned for the past and most recent gradients to represent their higher importance.

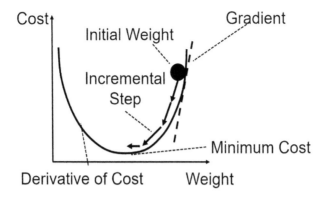

FIGURE 6.8 Locations of gradient descent.

- Weight updating process with adaptive learning rate

When the average value of the gradients for the most recent iterations is closer to zero, the model becomes an approximately flat surface. In order to converge and lead to a global minimum, a downward slope that has the effect of acceleration should be found. Here, the learning rate component should be increased to learn fast, when the gradient value is low. The inverse proportion connection is established by dividing the fixed learning rate by the average gradient value. When there is a high adapted learning rate, it results in a large weight update, as it is multiplied by the gradient. In contrast, when there is a high average gradient value, the model training is on steep slopes. In order to reach the global optimal point, need to take small steps by performing the same division and moving with caution.

- Momentum

Momentum is used as accelerating learning, especially with small, high curving, and consistent or noisy gradients. This accumulates the exponentially decaying moving average of recent gradients and moves in their directions. As shown in Figure 6.9, consider a scenario with some noisy data. Even though these dots seem close to each other, they do not share x coordinates. In such situations, a moving average that denoises the data should be used and brings data points closer to the original function. For instance, when the hyperparameter has a smaller number, the sequence will fluctuate, since smaller numbers of samples are averaged, and it will be closer to noisy data. When the hyperparameter has a large value, the curve will be smoother but shifted to the right, as it averages over a large number of samples. Thus, a value between these two extremes should be selected to obtain a balanced value for the hyperparameter.

6.6.2 GRADIENT-BASED OPTIMIZATION ALGORITHMS

Let us discuss some of the gradient-based optimization techniques used in machine learning.

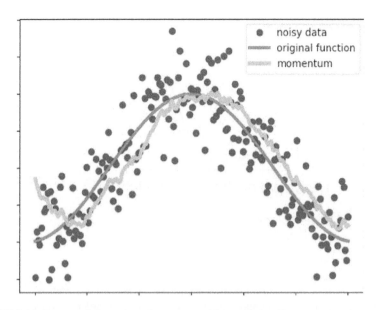

FIGURE 6.9 Exponentially weighted averages with a smoother line representation.

- Gradient descent (GD) optimizes a learning model by considering the learning rate and the current gradient and traversing the search space. It identifies the parameter values of function that minimizes the cost. It processes the entire dataset at once, thus resulting in quick convergence. However, it requires more resources to process the entire dataset at once. Gradient descent finds the local minimum of a differentiable function as shown in Figure 6.10. When there are many data points, it requires more time to produce the optimal vector over a set of iterations. In simple machine learning models, such as linear regression models, the gradient descent considers the entire dataset and calculates the loss during forward propagation. Then it updates the weights during backpropagation. This process requires more resources and extensive computations.
- Stochastic gradient descent (SGD) uses only a subset of data points to identify the local minima. Therefore, it addresses the issue with the gradient descent method by consuming less time. However, since it considers one record at a time and does forward and backward propagation in linear regression models, the training becomes very slow. Thus, taking more time to converge. SGD optimizers update the learning rate by taking the product of the learning rate and a factor based on the gradients.
- Mini-batch gradient descent: generally, when the dataset is large, the entire dataset is not feasible to train at once as it requires more resources (which reflects GD). Also, it may not be possible to load only one data point at a time, as it takes time (which reflects SGD). Thus, mini-batch SGD considers small random sets of instances at a time (where $k<$ learning rate), that will converge via a zig-zag path as shown in Figure 6.4. It takes a considerable among of time to converge, but it is a widely used method for large datasets. This

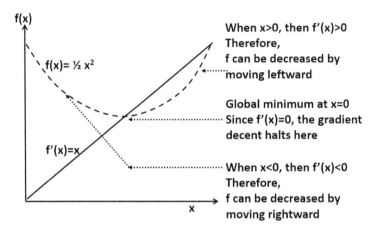

FIGURE 6.10   Representation of gradient descent.

boosts performance by using high-performance computers such as GPUs. The mini-batch SGD uses forward and backward propagation with mini-batches. Although this requires fewer resources and less time, it is affected by the noise.

- SGD with momentum is a popular optimization algorithm that is used to reduce the noise with mini-batches and train efficiently. Momentum or SGD with momentum accelerates gradient vectors in the forward directions resulting in efficient convergence. It uses the exponential moving average (EMA) to remove the noise and smooth the curve. As we discussed above, EMA assigns a high weight and importance to the most recent data points in different time intervals and measures trend direction over a period. This weighs the number of observations and uses their average. SGD with momentum also avoids local minimum when the process does not use batch normalization.

It is interesting to see the comparison of gradient-based methods as shown in Figure 6.11. The batch gradient descent calculates the gradient for each step utilizing the entire training dataset. Therefore, when the training dataset is large, this process becomes slow. The SGD considers an arbitrary data point in the training set for each step and generates the gradient. Since only one data point is required to store in memory at each iteration, the process performs efficiently. This is possible to apply to large datasets. However, the stochastic nature of SGD is not widely used compared to the batch gradient descent method. In SGD, the cost function fluctuates by reducing only on the average, without reducing gradually up to the minimum. Although it reaches near the minima, it continues to fluctuate without reaching a stable point. The fluctuations will prevent local optimal points and the final values perform better, however, it may not be the optimal solution. A learning scheduler can be used to address this, which reduces the steps while approaching the global minima.

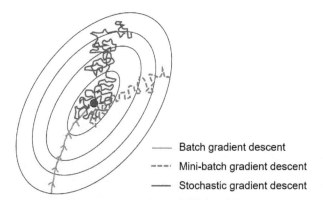

Batch gradient descent
Mini-batch gradient descent
Stochastic gradient descent

FIGURE 6.11   Convergence of gradient descent methods.

6.6.3   OTHER OPTIMIZATION ALGORITHMS

Other optimization algorithms such as Adagrad, Adadelta, and RMSprop optimize more compared to the above optimizers. Adam is a widely used optimizer that consists of a combination of SGD with Momentum and RMSprop.

- Adagrad (Adaptive gradient)

Adagrad uses parameter-specific learning rates, where the learning rate adjusts based on the rate of parameter updates while model training. The idea behind this technique is to use different learning rates for each hidden layer, and each neuron for different iterations or epochs. This process divides the learning rate by the square root of the collective sum of present and earlier squared gradients. As in SGD, the gradient will not change. This technique supports increasing the performance of sparse gradient tasks as it retains a per-parameter learning rate. Thus, mainly computer vision and NLP tasks use Adagrad. However, when the number of iterations increases, it will get a very small learning rate and will decrease the weight updating. Thus, there will not be a significant value to update the weights. Hence it delays the convergence. This problem can be addressed using AdaDelta and RMSProp optimizers.

- AdaDelta (adaptive delta)

Adadelta is an extension of AdaGrad optimizer and an SGD method. The term delta denotes the difference between two consecutive weight updates. This uses adaptive learning rate per dimension as a solution for the limitations namely, the repeated decline of learning rates during training and the requirement of manual selection of the global learning rate. Instead of using the learning rate as a parameter, AdaDelta utilizes the frequency of parameter change to adapt the learning rate. Additionally, this technique retains the second moments of gradient and the change in parameters as variables.

- RMSprop optimizer (root mean square propagation)

RMSprop is an extension of AdaGrad optimizer and is based on the adaptive learning rate. This is based on gradient descent with momentum. The declined moving average of partial gradients is utilized to adjust the update on each parameter. Thus, RMSprop considers the recent partial gradient values over the search space, without relying on previous gradients. Thereby addressing the limitation of AdaGrad. RMSprop takes the EMA of these gradients, without using the cumulative sum of squared gradients as in AdaGrad. As in momentum, this has become a typical update for the learning rate in many optimizers. Here, to avoid the summation of all derivatives of weights becoming a high or very small value, that prevents very small or high learning rates, respectively, these methods consider weighted average instead of derivative of weights. The per-parameter learning rate, which is based on the average of the values of the gradients for the weights in the recent past, is utilized by considering the changing frequency.

Since it prevents vertical fluctuations, a high leaning rate can be used, which results in taking large steps horizontally and converting efficiently. However, there is a chance of getting divergent behaviors due to drastic updates with very large learning rates. Figure 6.12 shows a representation of fast and slow learning rates.

When we compare the above three techniques, Adadelta uses EMA instead of the learning rate. AdaGrad and RMSProp utilize the second moment without and with decay, respectively. Thus, RMSProp speedups the process compared to AdaGrad.

- ADAM optimizer (adaptive moment estimation)

Adam is a widely used optimizer and different to classical SGD, which uses a fixed learning rate for a given model to update weights throughout the training. This uses the features of both RMSProp and Momentum. For instance, for the gradient, the EMA of gradients is used as in Momentum. To decide the learning rate of parameters, it divides the learning rate by the square root of the EMA of squared gradients as in RMSProp. The RMSProp optimizer adapts the learning rates of the parameters by

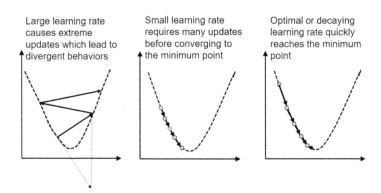

FIGURE 6.12   Representation of the behavior of different learning rates.

considering the mean, which is the average first moment. Whereas Adam uses the uncentered variance, which is the average of the second moments of the gradients. This allows for the management of sparse gradients on noisy tasks. Therefore, Adam processes the EMA of the gradient and the squared gradient by enabling it to handle the decay rates of the moving averages.

Generally, the performance of the Adam optimizer is good and used as the default optimizer for most of the applications due to the following reasons.

- Straightforward to implement.
- Low memory consumption.
- Computationally efficient.
- Faster computation time
- Requires few parameter tuning.
- Applicable for large tasks.
- Can be used for tasks with noisy or sparse gradients.

6.7   NEURAL ARCHITECTURE SEARCH (NAS)

6.7.1   Overview of NAS

Deep learning models have become autonomous and essential in many industries due to their significant accuracy and robustness. Generally, the design and development of an efficient neural network require the knowledge of architectural engineering, domain expertise, and time to explore an iterative process for the full range of solutions. Designing a neural network is not necessarily straightforward due to the aspects, such as intractable design space, non-transferable optimality, and inconsistent efficiency matrices that affect the development of deep learning models. For instance, neural network designers are faced with a combinatorically massive design space, where different possible architectures can be obtained by merely changing parameters, such as kernel size and the number of filters. When we include additional features, such as the number of layers and skip connections between layers, this creates a complex search space that is intractable to search manually and consumes more time. As a result, usually designers select a model with similar previous tasks and try to evolve the selected architecture in a few iterations. This approach is not sufficient to fully explore the potential architectural space.

In order to generalize neural networks while avoiding overfitting the training datasets, it is important to find optimized architectures. However, based on the objectives of the organization, where productivity is more important than quality and due to the lack of time and expertise, the efficiency of the models may not consider and proceed with a sufficient model that performs the task. Thus, based on the application domain, different architecture-specific solutions should be designed. Since most of the model designs depend on expert knowledge, there is a requirement to find an autonomous way to design and develop efficient and optimized deep learning models.

Neural architecture search (NAS) is used to automatically design optimal and efficient neural networks with low losses, for a given dataset. This method evaluates many deep learning architectures across a search space using a search strategy and selects the best match for the objectives of a given problem by maximizing a fitness function. This can be considered a biologically inspired optimization-based algorithm. NAS automates both the architecture selection and neural network training, which along with the increase in computational resources available should eventually allow the development of tailor-made architectures for each task per each hardware platform. NAS is used in various tasks, such as object detection and image classification in image processing, hyperparameter optimization and meta-learning in AutoML. NAS will bring flexibility to industries with deep learning applications that can adapt to diverse requirements. Thus, the advantages can be listed as follows.

- A formal algorithmic approach for developing neural networks.
- Makes model development easier for each hardware solutions.
- Can develop models highly competitive with models developed by human experts.
- Easier to regulate resource constraints.

Generally, NAS is not widely applied in real-world applications due to a few limitations. NAS methods explore many potential solutions with variable complexities. Hence it is computationally expensive in terms of resources and time. The larger their search spaces, the more there are architectures to test, train, and evaluate. Since the architectures are evaluated with training data, it can be hard to predict the performance on real data. Also, expert knowledge is required to speedup the search process by fine-tuning different architectures and guiding the search to converge quickly towards an optimal solution. Additionally, it may not produce the global optimal solution even with sufficient resources. Further, many NAS studies are irreproducible and may not be able to compare with the baseline models.

6.7.2 NAS Process

We have learnt that recurrent neural networks (RNNs) support the processing of sequential data to output the next possible element in the sequence. Since RNNs are prone to vanishing gradient and exploding gradient problems, LSTMs are applied by handling the importance of each of the prior data in the sequence by using different gates. Generally, NAS uses RNNs to handle sequential data, by decoding the sequential outputs to produce more suitable models iteratively. The controllers are designed to navigate the search space more intelligently using convolutional blocks and stacking some of them to design the learning model. A search space with basic convolutions, depth-separable convolutions with different kernel sizes and pooling layers is used to design cells. Here, the convolutional cells that output a feature map of the same size and half-size of the input are defined as normal cells and reduction cells, respectively.

The NAS can be explained in three terms as follows.

1. Search space: set of neural network architectures that solve a given problem.
2. Search strategy: the mechanism used to explore the defined search space for candidate architectures.

3. Performance strategy: evaluate the effectiveness of the architectures that are generalizable or perform well in unseen data.

In the NAS process, a search strategy selects a model from the search space and evaluates the selected architecture's performance and passes it to the search strategy. NAS predicts hyperparameters, such as the height and width of filters and strides, the number of filters, and skip connections. Each prediction is performed using a Softmax activation and passed to the next layer. Additionally, NAS models have been developed with reinforcement learning. Here, also an RNN is used as the controller to sample the search space. After training and evaluating the performance of the sampled architecture, the result is utilized to update the controller using gradient-based methods and the process is iterated until convergence or timeout.

The process of NAS is shown in Figure 6.13 and can be listed as follows.

1. Process the search space consists of possible models with predefined operations such as convolutional, fully connected, pooling, and recurrent layers and their connections.
2. Use a controller to select possible architecture from the search space.
3. Train the selected architecture for a few epochs
4. Evaluate the candidate model and rank them by considering the performance on the validation set.
5. Use the obtained ranking to rearrange the search and update the controller accordingly to identify new possible models.
6. Repeat the procedure until the optimal architecture is found to preserve a given condition.
7. Evaluate the best model on the test dataset.

Neural networks discard some features from the model based on their importance to make decisions. The NAS process discards features from the search pipeline training with hidden layers to identify their importance. Further, it is important to consider model compression when designing NAS. Approaches, such as quantization, pruning, and knowledge distillation, have been used to compress the models.

6.7.3 SEARCH SPACE

This defines a set of feasible solutions for a NAS method. Search space can be configured with the properties gained from prior knowledge of the type of problem

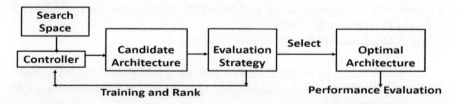

FIGURE 6.13 Process of NAS.

to be solved. Thus, although this may lead to human bias, it can simplify the search process and deliver timely results. NAS includes a reinforcement learning-based approach, which is in the discrete space and a gradient-based approach in the continuous space.

The search space can be considered in different types as follows.

- Global search space

This covers graphs that represent an entire neural architecture. This search for all possible combinations of operations results in an expensive search space. It combines operations to form a chain or sequence of architectures. The associated parameters are the number of layers, operation type, and the associated hyperparameters. Skip connections are also added to allow multi-branch architectures.

- Cell search space/modular search space

The search space is modularized by combining a set of modules that consists of various layers and blocks. The modularization via cells discovers the cell architecture that can be merged to create the neural model as in ENAS (efficient NAS). Also, the block diversity can be emphasized by changing the optimizing blocks while holding the remaining blocks fixed as in FPNAS (fast and practical NAS). Further, a layer-based search space can be utilized as in FBNet (Facebook-Berkeley-Nets). Here, a unique block can be selected by each searchable layer across the search space.

- Stochastic NAS (SNAS) space

A collection of one-hot random variables from a fully factorizable joint distribution make up stochastic NAS (SNAS) space. By loosening the architecture distribution with concrete distribution, gradient optimization is made possible, making this search space differentiable.

Generally, we assume that the possible architectures are in a discrete space. Differentiable architecture search (DART) is utilized to convert the space into a continuous and differentiable form. It represents each node in a directed acyclic graph (DAG) to convert the module-based space into a continuous space. Here, each DAG has two input nodes and one output node and is created by sequentially connecting N nodes. The transformation is performed by representing the discrete set of operations as a Softmax that gives a sequential set of probabilities. As another method, neural architecture optimization (NAO) can be used. Here, by mapping a discrete search space to a continuous space, an optimal embedded encoding can be obtained utilizing gradient optimization. The best continuous representation is then discretized into the final architecture using a decoder. These one-short techniques, DARTS and NAO, use supermodels to determine the ideal designs.

The search space can be designed in different architectures as shown in Figure 6.14. For instance, chained-structured neural networks have convolution and pooling layers chained up. Modern implementations use more elements such as separable

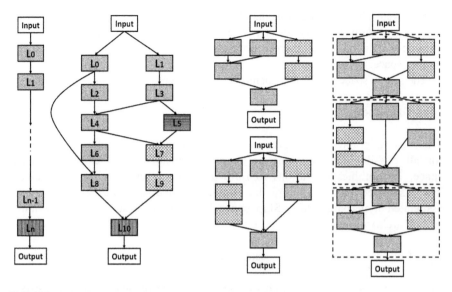

FIGURE 6.14   Sample architectures in the search space.

convolutions and dilated convolutions where the parameters determine the conditional space. Multi-branch networks and cell-based search are widely used in image classification applications.

The search space can be either coarsely granular, where a block consists of many layers, or more finely granular, where one layer with a given kernel size. The architectural search is more adaptable with the finer-grained procedures, enabling the generation of more complex designs. However, the search results will take longer to converge as a result, and preliminary results might not be ideal. As long as the search strategy starts with a sound initial architecture, a coarser search space can leverage 'neural blocks' based on the domain's current knowledge to get better predictions with fewer iterations. However, in the long term, applying finer-granular search space is more flexible than the neural blocks, as the finer-granular process can independently find such neural blocks and subsequently enhance them.

6.7.4   SEARCH STRATEGIES OF NAS

Recent efforts in designing neural networks do not solely focus on accuracy but also aim to optimize efficiency such as implementing them on devices with limited memory capacities. When additional features such as the number of layers and skip connections are included between layers, this creates a combinatorically massive design space, and the search space becomes intractable to search manually. Additionally, the optimality of architecture depends on the specifications or the configurations of the target device, such as power, energy, memory capacity, and latency. Further, the same architecture may not perform equally well across different devices and platforms. This can be addressed by neural architecture search (NAS) strategies that explore the search space to find well-performing architectures quickly while avoiding premature convergence that leads to overfitting.

The methodologies that can be used in this step include Bayesian optimization, random optimization, evolution methods, reinforcement learning, and gradient-based methods. Other methods such as neuro-evolution, network morphing, and game theory can be used. The reinforcement learning algorithms and evolutionary algorithms have shown higher accuracy values in terms of test accuracy with image classification. Furthermore, data augmentation, depth, and the number of filters affect positively the accuracy of test data. The process utilized to find the ideal architecture inside the search area is referred to as the search strategy. The areas of NAS algorithms based on the search strategy can be listed as follows.

- Grid search: the simplest form of search strategy where the screening is taken place systematically screening.
- Sequential model-based optimization: select a model by iteratively exploring the suitable models.
- Random search: select architectures at random from the search space and test them according to the performance estimation method.
- Bayesian optimization: useful in hyperparameter search. Uses probability distribution and observations from tested architectures.
- Evolutionary or genetic algorithms: existing models may recreate additional architectures or eliminate them from the search space.
- Reinforcement learning.
- One-shot architecture search: trains a supermodel that includes all other configurations in the search space using parameter sharing.

The search strategies can categorize into black-box optimization strategies or differentiable architecture search strategies.

1. Black-box optimization strategies.

This operates in a discrete space, making it impossible to directly optimize with methods like gradient descent. However, by utilizing techniques like evolutionary programming, reinforcement learning and Bayesian optimization, we can optimize across desecrated space. While using different optimization techniques for these tactics, like restricting the search space to adhere to specific structures, weight sharing, and performance prediction, they require a more computational cost. The black-box optimization algorithms are expensive to deploy in the majority of real-world use cases since they need many architecture assessments and each evaluation needs partial or full training of a neural network.

2. Differentiable architecture search (DARTS)

This is a method for efficient architecture search that makes a continuous search space, enabling it to be optimized using gradient descent. This combines both the search and evaluation stage into one. Compared with black-box optimization strategies, the main advantage of gradient descent is its data efficiency allowing it to achieve competitive results with fewer iterations.

The search space is represented by DARTS as a directed acyclic graph including input and output nodes. A tensor is a node, and an operator is an edge-linking node. Each node has a connection to every prior node. The categorical option is loosened into a Softmax over all feasible procedures. As a result, the architecture supports a continuous search space and refinement in a single network using a bi-level optimization technique for both weights and architecture.

Here, gradient descent is utilized to learn the parameters and weights. The final configuration is obtained by applying argmax on each edge. Comparing this search approach to black-box optimization strategies, it performs better. However, DARTS consume more memory with the increase of the candidate set size. Figure 6.15 shows the operations in DARTS, where (a) shows the undefined operations along the edges, (b) shows the continuous relaxation of the search space, (c) represents the combining probabilities and weights for optimization, and (d) shows the final architecture generation.

On the other hand, memory consumption can be addressed by techniques such as ProxylessNAS. It uses path-level binarization to reduce the memory consumption of differentiable search strategies. Also, latency in the target device can be used as device sensitive performance metric to optimize the architecture for each target device. FBNet and PC-DARTS are another two developments based on DARTS.

- FBNet

FBNet is a type of convolutional neural architecture discovered through DARTS. It is a blockwise differentiable NAS method (blockwise DNAS), where the best

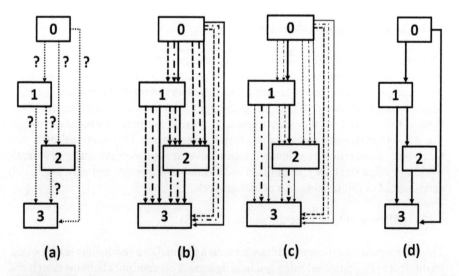

FIGURE 6.15   Representation of DARTS.

candidate building blocks can be chosen by using Gumbel Softmax random sampling and differentiable training. It also used latency to optimize search results for each target device. FBNet utilizes a basic type of image model block inspired by MobileNetv2 that utilizes depthwise convolutions and an inverted residual structure. At each searchable layer, the diverse candidate blocks are side by side planned, leading to sufficient pretraining of the supernet. The pretrained supernet is further sampled for fine-tuning of the subnet, to achieve better performance. FBNet uses cross-entropy loss to lead to better accuracy and latency loss to optimize the search results on a target device.

- PC-DARTS

Partially connected DARTS is another development of DARTS, that uses partial channel connections, which reduced memory consumption and computational cost. To eliminate network space redundancy and execute an efficient search without sacrificing performance, it samples a subset of the super-model. PC-DARTS is trained with a large dataset leading to efficient, fast and high training stability, because of the low memory requirement. Complicated network architectures such as skip connections can be easily discovered with PC-DARTS, unlike in the case of FBNet.

Unlike in DARTS, where all the channels are sent to the operation selection, PC-DARTS architecture samples the channels by a factor of K, such that only a $1/K$ proportion of channels are selected for the operation selection. Here, K is a hyperparameter and can be varied to balance the trade-off between the accuracy and the efficiency of the architecture search. Thus, it reduces the memory and the computational overhead of the architecture search by a factor of K. Therefore, it provides an additional advantage of the ability to train a large dataset and make the overall training faster. However, on the downside of using partial channel connections, these architectures may suffer from significant instability. Thus, PC-DARTS used an edge normalization technique to contribute to the stability of the architecture search.

For instance, this can be used as a hardware-agnostic NAS that considers only the input and output for the optimization, and thus does not depend on the hardware specification. The memory consumption of these models was kept within the limits of a typical edge device can handle. These models may use complex computations to achieve better accuracy, however, it may increase the latency.

6.7.5 Strategies for Performance Measures

The performance of the derived model architectures can be measured mainly using the test accuracy, training time, and computational cost. However, these metrics can be varied based on the problem and the application domain. For instance, medical data classification should focus more on specificity and sensitivity rather than accuracy itself. Also, the computational cost can be taken for granted on some occasions based on the task complexity. Also, it is important to perform the performance estimation of NAS methods efficiently.

Some of the performance estimation strategy methods can be listed as follows. Also, this can be used to reduce the computational time associated with the NAS process.

- Validation accuracy: although this is a simple method, it may require more time and more computations in scenarios such as large datasets, large search space, and models with more layers. This can be addressed by strategies such as low-reliability measures achieved by model training with a small number of epochs or a subset of data.
- Measurement of low reliability: use a small set of epochs or data to train and reduce the model size.
- The exploitation of the learning curves: this can be used to measure the model performance without completing the training cycle. The efficiency of the search process can be increased by discarding the models that produce low predictions during the first few epochs.
- Efficient NAS method: uses subgraphs to explore the search space yields better results and reduces the computational cost.
- Weight sharing one-shot models: the sub-models use the weights from the supermodel.
- Train a substitute model to predict the model performance based on characteristics derived from other new architectures.
- Training with warm-start: use the supermodel weights for the weight initialization.

6.8   ADVERSARIAL TRAINING

6.8.1   OVERVIEW OF ADVERSARIAL TRAINING

At present, there are many support tools, libraries, and online services to apply deep learning capabilities to applications easily without a thorough knowledge of machine learning. As for many systems, these models can have the threat of adversarial attacks, which is an important concern of machine learning applications. Adversarial machine learning aims to mislead the learning process by giving misleading input. It addresses the creation of attacks on machine learning algorithms and the detection of adversarial examples, providing defensive mechanisms against such attacks. Adversarial attacks are mainly found in spam detection and image classification tasks.

Contrary to other security dangers that programmers are accustomed to dealing with, adversarial attacks are unique. The traditional cybersecurity environment has developed to handle a variety of software threats. There are many static and dynamic analysis tools to identify and fix security bugs in software. Compilers detect and identify unsuitable and possibly harmful source code. Unit testing ensures the responses of the functions to various inputs. The browser and the computer's hard disk can both be searched for and blocked by anti-malware software and other endpoint solutions. Web application firewalls can check and deny malicious requests to web servers. Code and app hosting platforms are also in-built with security applications.

Compared to general cyber-attacks, the nature of attacks in deep learning models is different. These threats try to exploit and modify the behaviors of deep learning models. Although many threats affect image classification such as mislabeling images, these can occur in text and audio as well. Generally, when the accuracy of a model increases, the adversarial robustness decreases. Therefore, the requirement is to develop models that are both accurate and robust against adversarial attacks, avoiding a trade-off between accuracy and robustness.

Deep learning models are applied in sensitive and dependable applications in autonomous driving, finance, and healthcare and the associated adversarial attacks indicate the differentiation of the decision-making processes of humans and learning models. It affects the trustworthiness of the deep learning model. For instance, adversarial attacks in safety-critical applications can affect negatively the lives and health of those who will employ the machine learning models directly or indirectly in danger. It can violate people's rights in fields like finance and hiring while harming the reputation of the business using the models. Attackers can manipulate models in security systems to get around facial recognition and other ML-based authentication methods. Because the faults and assaults might be volatile, several businesses are hesitant to use them.

Different model visibility levels are considered in deep learning security. In a white-box threat, the attacker has total control over the target model, including its architecture and parameters. For instance, a model that is published online (GitHub) is considered a white-box model as there is direct visibility of the architecture and parameters of the model to the outside. Thus, it is more prone to adversarial attacks. A black-box attack is a situation in which the attacker is unable to access the model and is only able to see the model's outputs. Consider a scenario, where online APIs such as Amazon Recognition or Google Cloud Vision is used to access a model. Here, the model is visible as a black-box, where the outsiders have access to the output of the model only. Thus, is it harder to attack, however, model-agnostic adversarial attacks can be applied to learning models with black-box nature.

6.8.2 Types of Adversarial Attacks

Different types of malicious activity can happen at different stages of the deep learning process. Some attacks are intended to compromise a learning model's integrity, causing it to output erroneous data or result in a specific outcome that the attacker planned. Other adversarial assaults might target a system's privacy and force the model to disclose sensitive or private data. Let us examine the various adversarial attack types and the machine learning pipeline's flaws.

- Data poisoning (contaminating) threats: the attacker adds falsified data to the training set. An adversary tricks a classifier into making erroneous or biased conclusions by feeding it wrongly labeled data and identifying the peculiarities of the data incorrectly. Here, during the training phase, the misleading data is used to fine-tune the model parameters and make it sensitive to adversarial deviations. During testing, a misclassified model will act unpredictably.

TABLE 6.1
Types of Adversarial Attacks

	Data Poisoning	Backdoor Attacks	Membership Inference	Model Extraction	Model Evasion
Original data	x	x			
Model training	x*				x*
Model testing			x	x	x

- Backdoor attacks: this is an extension of data poisoning. When an adversary inserts visual patterns into a training dataset, this is known as data poisoning. Then, the attacker uses such patterns during the testing phase to cause the target deep learning model to act in a certain way.
- Membership inference attacks: this is an inference-time attack that tries to reveal the training data and obtain sensitive data from the target model. For instance, the extraction of credit card numbers or passwords can be considered.
- Model extraction (stealing): here, an attacker explores a black-box model to reconfigure the model or recover the data it was trained on. This is important in processing sensitive and confidential data and models. For instance, the adversary could get financial benefits by stealing a stock market prediction model.
- Model evasion: a popular adversarial example that starts with a normal input and progressively adds noise to increase the biases of output toward the expected outcome such that obtaining a specific output or decreasing the confidence of the prediction. During deployment, the attacker manipulates the data to mislead classifiers that have already been trained. This is the most practical type of attack, as it is performed during the deployment phase, and mostly uses intrusion, spam, or malware scenarios to avoid detection by complicating the content by bypassing the filters and classifying it as acceptable without directly affecting the training dataset. For instance, spoofing attacks against biometric verification systems can be considered.

Table 6.1 states the impact of each adversarial attack on the dataset, training, and testing phases. The symbol * denotes that there is no access to the internal information of the model and is considered a black-box attack setting. Figure 6.16 shows the process behind these attacks.

6.8.3 ADVERSARIAL ATTACK GENERATION TECHNIQUES

- The fast gradient sign method (FGSM) is a quick and easy gradient-based method. It is used to create adversarial tasks to reduce the maximum amount of perturbation that may be applied to any individual pixel of the image to lead to misclassification. FGSM attack is a white-box attack with the goal of misclassification.

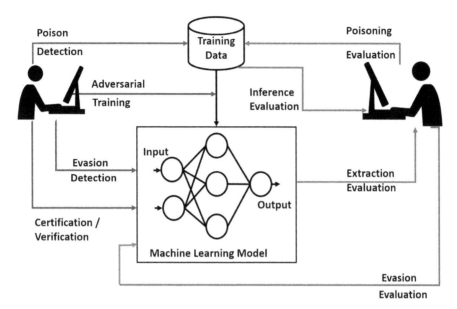

FIGURE 6.16 Adversarial attacks.

Advantage: relatively fast computation time.

Disadvantage: rach feature is added with perturbations.

- Jacobian-based saliency map attack (JSMA) is a feature selection that helps to reduce the number of features that need to update while producing erroneous predictions. Flat perturbations are systematically applied to features in descending order of saliency value. The goal of this greedy method is to raise the predicted erroneous outputs over numerous iterations that modify one pixel at a time.

  Advantage: perturb few features.

  Disadvantage: computationally complex compared to FGSM.

- Deepfool attack is an adversarial sample generation method without a target. It reduces the Euclidean distance between perturbed instances and the original data. It calculates the decision boundaries among classes and repeatedly introduced perturbations.

  Advantages: produces adversarial instances well, with higher misclassification rates and lower perturbations.

  Disadvantages: requires more processing than FGSM and JSMA. Additionally, adversarial instances are probably not the best.

- Carlini & Wagner attack (C&W) relies on the Limited-memory Broyden-Fletcher-Goldfarb-Shanno (L-BFGS) threats in the optimization process. It excludes box limitations and uses other objective functions. As a result, the technique is better at producing adversarial examples. Baseline defeats like defensive distillation and adversarial training are defeated by it.

Advantages: generate adversarial elements effectively and can overcome selected adversarial defenses.

Disadvantages: computationally complex compared to FGSM, JSMA, and Deepfool.

- Generative adversarial networks (GANs) utilize two neural models to produce adversarial threats. The result is that one behaves as a generator and the other as a discriminator. The generator aims to provide samples that the discriminator will misclassify in a zero-sum competition between the two networks. The discriminator makes an effort to discriminate between genuine samples and those produced by the generator.

Advantages: produce instances that are different from a training set.

Disadvantages: computationally complex and unstable.

- The zeroth-order optimization attack (ZOO) method is perfect for black-box attacks since it enables the calculation of the gradient of the classifiers without having access to the model. By querying the expected model with updated individual features, the method calculates gradient and hessian and applies Adam's or Newton's method to optimize perturbations.

Advantages: performance is the same as the C&W attack. Does not require training in the alternative models or details about the classification model.

Disadvantages: need many queries to obtain the expected output.

6.8.4 Adversarial Attack Defensive Methods

Detecting adversarial attacks is challenging due to the black-box nature of deep learning models, where the actual cause cannot be identified correctly. Some of the solutions can be listed as follows.

- Adversarial training: the system is trained to learn the possible adversarial attacks, simulating the design of an immune system. This will improve the robustness of the model to produce a set of threats against a system beforehand. However, this method may not avoid all types of threats, as it is impossible to produce all conceivable attacks in advance due to their size.
- Continually alter the learning algorithms: by keeping the algorithms confidential and rarely modifying the model, a moving target can be produced. One option is to extract and save the data at the time of acquisition and then assess it with the input of the algorithm.
- Limited-memory Broyden-Fletcher-Goldfarb-Shanno (L-BFGS) is a non-linear gradient-based numerical optimization methodology. It reduces the number of perturbations that are added to images. This requires less computer memory and is used for parameter estimation in machine learning.

Advantages: performs better in producing adversarial elements.

Disadvantages: computationally complex, consumes more time and is difficult to use in practice, as the optimization is based on box constraints.

6.8.5   BEST PRACTICES TO AVOID ADVERSARIAL ATTACKS

- Select the dataset from a trustworthy resource, which is evaluated by a recognized organization. Generally, the datasets that are extensively used in deep learning applications have a higher integrity. The datasets from unknown sources may contain hidden malicious patterns that can poison the model training.
- Take precautions to anonymize sensitive data such as credit card details. Attackers may use membership inference techniques to explore the training data, even if the data is kept private.
- Hire a trustworthy developer to develop a deep learning model that has not been harmful or is not susceptible to adversarial attacks.
- Check the truthfulness of the provider before downloading a deep learning algorithm from open-source sites such as GitHub or PyTorch Hub and incorporating it with the application. If the publisher of the source code is a reputed organization, then the model may not have been intentionally poisoned or adversarially compromised, however, it may have unintentional adversarial vulnerabilities.
- Awareness of user access when the model is open-source and publicly available. Potential attackers can conduct adversarial attacks on any other application that utilizes the same model out of the box since they have access to the same model.
- Set defenses to account for malicious behavior. Even if the model uses an industrial API, the same API can be used to create an adversarial model by the attackers. Methods such as passing the input images via different scaling and encoding types can be applied to neutralize the adversarial perturbations.
- Restrict access to the deep learning pipeline. Allows the accessibility of the training data and model backend to trusted developers. When the training data is taken from an outside resource such as the comments, feedback, images, and reviews given by the clients, use a mechanism to avoid the in-floor of malicious data in the learning process.
- When there is an attack, roll back the model to a previous state, using a versioning system.
- Experience the latest tools to detect adversarial vulnerabilities. For instance, the open-source Python package called IBM's Adversarial Robustness Toolbox offers a set of methods to assess the learning models against various forms of threats. Another open-source program that scans learning models for malicious flaws is Microsoft's Counterfit.

REVIEW QUESTIONS

1. Explain how we can apply data augmentation for non-image data. Consider data types, such as tabular data, and multimodality imaging data such as MRI, fMRI, etc.
2. Explain the importance of normalization and how we can improve the model performance by normalizing the dataset.

3. What are described as hyperparameters? What techniques are used to fine-tune hyperparameter values?
4. What is the function of optimizers, come up with an optimizer of your own and describe with the code how it works. (Hint: you may use available mathematical optimizers.)
5. What do you see as improvements in neural architecture search in the future?

7 Performance Evaluation Techniques

7.1 OVERVIEW OF PERFORMANCE MEASURES

Assessing deep learning models is important for any application and this chapter discusses the associated performance metrics. Generally, accuracy is the most frequently used model evaluation technique. When the data is skewed, accuracy may be deceptive. In such situations, additional techniques like precision, recall, specificity sensitivity, receiver operating characteristics (ROC), and area under the curve (AUC) should be examined. It is essential to utilize different evaluation metrics on a given algorithm because a model may behave well with a specific metric but perform less with another evaluation metric.

The ground truth in deep learning refers to the real target expected for training or validating the model with a dataset. Although empirical research has shown that deciding the metric to employ for diverse situations is challenging, each has unique characteristics that assess different elements of the algorithms being examined. Due to significant weighted differences that often occur between projected and real value or otherwise, it is sometimes difficult to declare the best metrics for evaluating algorithms in specific application domains.

Let us learn the fundamentals of these metrics and how they operate. Classification applications rely on four main outcomes to generate this data. Here, FP and FN are considered as type I error and type II error, respectively.

- True positive (TP): both the ground truth and the predicted output are positive. True positive is the correctly predicted positive value, which indicates the number of classes with actual output is true and the predicted output is true, as well. As an example, suppose we have classification problems to identify whether a patient has a given disease or not. The true positive value will be the number of patients who have the disease (positive) and predicted as patients with the disease (positive).
- False positive (FP): the ground truth is negative, and the predicted class is positive.
  False positive is the class determines as true when it is actually false (type I error). It is considered as a 'false alarm'. For example, consider a scenario

DOI: 10.1201/9781003390824-7

where the number of patients who are healthy (negative) but predicted as patients with the disease (positive).

- True negative (TN): both the ground truth and the predicted output are negative. True negative is the correctly predicted negative value, which indicates the number of classes with the actual output is false and the predicted output is false, too. For example, the number of patients who are healthy (negative) and the predicted class tells the same thing (negative).
- False negative (FN): the ground truth is positive, and the predicted class is negative.

  A false negative is predicting a class as false when it is actually true (type II error). For example, the number of patients that were diagnosed with the disease (positive) but predicted as healthy (negative).

Deep learning performance assessments include some level of trade-off when characteristics that enhance one aspect of performance decrease another type of performance. We will discuss the trade-offs later in this section.

7.2 TYPES OF PERFORMANCE METRICS

7.2.1 CONFUSION MATRIX

The confusion matrix records the correctly and wrongly identified samples for each class and supports obtaining many other performance metrics. A confusion matrix can be defined for the n-class classification problems. Figure 7.1 shows four possible outputs that correspond to the elements of the 2×2 confusion matrix. Confusion matrixes are critical for understanding the model's granular level efficiency and determining whether the model is good, depending on the sensitivity of the use case. For example, medical domain applications require the model to have a low FN rate.

7.2.2 ACCURACY

Accuracy is the most intuitive measure, which is denoted by the ratio between correctly predicted instances to the total number of instance as given in (7.1). This works

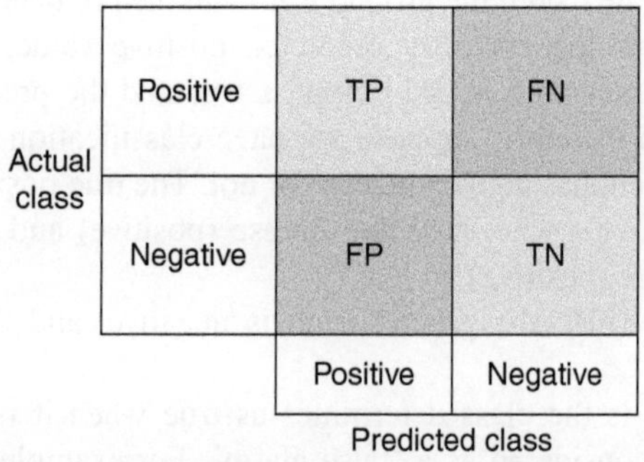

		TP	FN
	Positive	TP	FN
Actual class			
	Negative	FP	TN
		Positive	Negative

Predicted class

FIGURE 7.1   2×2 confusion matrix for binary classification.

well for symmetric datasets with almost the same number of FP and FN values. For datasets where the class distribution is not balanced, it is hard to differentiate the FP and FN values. Accordingly, it calculates the correct results' percentage that a classifier has achieved and indicates how well the classification model predicts the class labels specified in the problem statement. For binary classification, positive and negative concepts are utilized to measure accuracy.

$$\text{Accuracy} = \frac{\text{Number of correct predictions}}{\text{Number of total predictions}} = \frac{\text{TP} + \text{TN}}{\text{TP} + \text{FP} + \text{FN} + \text{TN}} \quad (7.1)$$

The term error rate (ERR) is defined as the complement of accuracy, which indicates no misclassified samples of the positive and negative classes, as in (7.2).

$$\text{EER} = 1 - \text{Accuracy} \quad (7.2)$$

Let us consider the following example that shows the model results that classified chest X-ray images as pneumonia (the positive class) or normal (the negative class).

$$\text{Accuracy} = \frac{1 + 90}{1 + 90 + 1 + 8} = 0.91$$

As shown in Example 7.1, the model accuracy is 91% (91 correct predictions out of 100 total samples). In other terms, out of 100 chest X-ray samples 91 are classified as normal (90 TNs and 1 FP) and nine samples are classified as pneumonia (one TP and eight FNs). Thus, the model could correctly identify 90 out of 91 normal subjects. But out of nine pneumonia cases, the model only correctly identified one as pneumonia, which is not good as eight out of nine pneumonia cases have been undiagnosed. Even though this model has good accuracy at first glance, another model that always predicts normal cases would have a similar accuracy on this set of samples. That means that this model does not perform any better than a model with no ability to forecast the difference between pneumonia and healthy participants.

EXAMPLE 7.1
Chest X-Ray Classification

True positive (TP)	False positive (FP)
- Reality: pneumonia	- Reality: normal
- Model prediction: pneumonia	- Model prediction: pneumonia
- Number of TP = 1	- Number of FP = 1
False negative (FN)	True negative (TN)
- Reality: pneumonia	- Reality: normal
- Model prediction: normal	- Model prediction: normal
- Number of FN = 8	- Number of TP = 90

Therefore, the metric accuracy alone does not reflect the entire performance, with a class imbalanced dataset, which has a considerable difference between the amount of positive and negative classes. Therefore, better metrics than accuracy such as precision and recall are needed to evaluate class-imbalance problems.

7.2.3 PRECISION AND RECALL

Precision and recall differentiate between proper label categorization within different groups.

Precision gives the proportion of a positive identification, which was actually correct. It counts the number of predictions from the positive class that are actually in that class. It is calculated as the proportion of accurate positive findings to those that the classifier anticipated to be positive as stated in (7.3).

Recall gives the proportion of actual positives that are identified correctly. It calculates how many positive class predictions were produced using all of the dataset's positive instances. It is calculated as the total number of true positive samples divided by the number of all samples that should have been identified as positive, as given in (7.4).

$$\text{Pr} \, ecision = \frac{\textbf{TP}}{\textbf{FP} + \textbf{TP}} \tag{7.3}$$

$$\text{Re} \, call = \frac{\textbf{TP}}{\textbf{TP} + \textbf{FN}} \tag{7.4}$$

While improving recall will reduce the amount of false negatives, maximizing precision will reduce the number of false positives. When minimizing false positives is the main goal, precision is appropriate.

Let us calculate the precision for Example 7.1. Chest X-ray classification. Here, we have TP = 1, FP = 1, FN = 8, and TP = 90. Therefore, $\text{Precision} = \dfrac{1}{1+1} = 0.5$ and $\text{Recall} = \dfrac{1}{1+8} = 0.11$

Precision counts the correct percentage of everything that has been anticipated to be positive.

Precision can interpret as follows.

- A noisy selection process allows a less accurate model to identify numerous positives. Additionally, it falsely detects a large number of positives that are not positives.
- A best model is extremely 'pure'. Although, we do not discover all of the positives, the ones that the model does identify as positive are almost certainly true.

Recall is defined as follows.

- How many instances are found to be truly positive by the model, among all the actual positive instances.
- Even if they might mistakenly classify some negative examples as positive, a model with high recall does a good job of discovering all the positive instances in the data.
- All or a significant portion of the positive instances in the data cannot be found by a model with low recall.

Precision and recall measures performance better, when the data is unbalanced, because they consider different types of errors (FP and FN) that the model generates. For a complete evaluation of the model, both precision and recall should be evaluated. However, improving precision typically causes reduced recall and vice versa. Because of this tension between precision and recall, some metrics such as the F1 score have been developed which rely on both metrics.

7.2.4 F-MEASURE

F-measure or F1-score offers a score that strikes a compromise between recall and precision concerns, considering their weighted average. It is a statistical metric used to evaluate performance. In other words, an F1-score is the average performance depending on the variables recall and precision and can be defined as in (7.5). As a result, the formula accounts for both FP and FN. The F1-score is in the [0, 1] range. This shows the reliability, so that will not miss the instances, and accuracy of the model.

$$\text{F1-score} = 2.\frac{\text{precision . recall}}{\text{precision + recall}} \qquad (7.5)$$

Generally, when the class distribution is uneven, accuracy will not be a good measurement. In such scenarios, it would be better to consider both precision and recall. Thus, the F1-score works better than accuracy to figure out how good the model performed. A very accurate result is produced by high accuracy and with low recall, however many occurrences that are challenging to identify are thus missed. The performance of the model improves with increasing F1-score. Thus, the class imbalance problems can be addressed by the performance metric F1-score, that based on the type and number of prediction errors.

Points to note:

A model will obtain a high F1-score if both precision and recall are high.
A model will obtain a low F1-score if both precision and recall are low.
A model will obtain a medium F1-score if one of precision and recall is low and the other is high.

7.2.5 SPECIFICITY AND SENSITIVITY

Sensitivity and specificity are frequently utilized in medical applications and models that use image and visual data. These metrics assess a classifier's performance on distinct classes independently. As an example, sensitivity is explained as the correctly recognizing patients with the disease. Specificity is described as the capacity to accurately identify those without the disease. A highly sensitive test produces fewer FN results and, leads to missing fewer instances with the disease. Sensitivity is the same as the recall in binary classification. Thus, sensitivity evaluates the capability of a model to predict TP for available categories. TPR denotes true positive rate as given in (7.6). Specificity assesses the capability of the model to predict the TN of each class. It behaves as the opposite of sensitivity. TNR stands for true negative rate, and it is the ratio of TNs to actual negative instances as given in (7.7).

$$\text{Sensitivity} = \frac{TP}{TP + FN} = \frac{TP}{P} = TPR \qquad (7.6)$$

$$\text{Specificity} = \frac{TN}{FP + TN} = \frac{TN}{N} = TNR \qquad (7.7)$$

7.2.6 RECEIVING OPERATING CHARACTERISTIC CURVE (ROC)

Receiving operating characteristic (ROC) curve is utilized to represent the performance of a binary classifier at all classification thresholds. It shows the classification results from the greatest positive classification to the most negative classification. The ROC curve is created by plotting the TPR against the false positive rate (FPR) at different thresholds. FPR is defined as in (7.8). Here, x-axis shows the FPR and y-axis represents the TPR. Figure 7.2 shows the ROC curve plot.

$$FPR = \frac{FP}{FP + TN} = \left(1 - \text{specificity}\right) \qquad (7.8)$$

A model needs to evaluate several iterations with numerous classification thresholds, to identify the points in the ROC curve; however, it is not efficient. This can be addressed by using area under the ROC curve (AUC), which is an efficient algorithm based on sorting.

- When the threshold is high or more like (0, 0), the specificity increases and sensitivity decreases.
- When the threshold becomes low or more like (1, 1), the specificity reduces and the sensitivity increases.
- The better performing test will have a higher region under AUC with the curve nearer to the upper left corner.

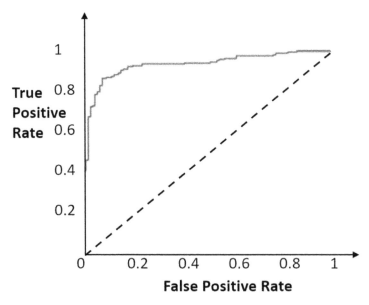

FIGURE 7.2 ROC curve.

7.2.7 AREA UNDER THE ROC CURVE (AUROC) AND AUC

The area under the curve (AUC) represents a single point on the ROC curve. It evaluates binary classification process. AUC is the same as the ROC diagram, and both will utilize the parallel order information. AUC gives a combined performance metric over the classification thresholds. As shown in Figure. 7.3, AUC measures the two-dimensional area under the ROC curve from (0,0) to (1, 1). A higher AUC value, the better the performance.

Useful features of AUC are as follows.

- Scale-invariant: indicates the appropriateness of the prediction rankings, instead of assessing the model predictions' absolute values.
- Classification threshold invariant: indicates the suitability of the predictions regardless of the selected classification threshold.

7.2.8 CROSS-VALIDATION

A deep learning model will be evaluated for effectiveness and accuracy in a real-world situation using a distinct and different dataset. Cross-validation is used to test whether a model behaves effective enough on the unseen test data compared to the training dataset. This assesses the model performance within the unknown data set. Thus, cross-validation can be used to indicate data underfitting or overfitting, and the model generalizability for the unseen dataset.

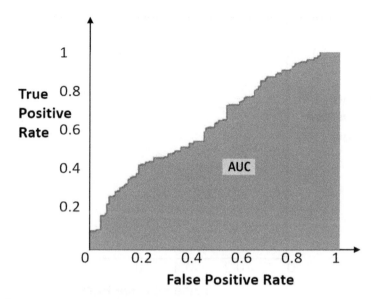

FIGURE 7.3 Area under the ROC curve.

Let us refresh our knowledge on the following terms.

- Training dataset: candidate algorithms are trained on this to fit for the model.
- Validation dataset: utilized to compare their performances and select the most performing model. In the process of fine-tuning model hyperparameters, it offers an unbiased assessment of a model's fit to the training dataset.
- Test dataset: accuracy, sensitivity, specificity, and F-measure are performance metrics that are obtained using the test dataset. It offers an objective assessment of how well a final model fits the training dataset.

The model is put to the test in every conceivable way throughout the thorough cross-validation. The initial dataset is split into training and validation sets to do this. Examples include leave-one-out cross-validation and leave-*p*-out cross-validation. Non-exhaustive cross-validation, such as the hold-out approach and *k*-fold cross-validation, does not separate out every potential combination and permutation from the original data set.

Cross-validation can be performed in several ways as follows.

- Hold-Out Cross-Validation

Hold-out cross-validation removes a subset of the training instances and perform predictions for the remaining instances. The obtained error prediction indicates the model performance on new instances or the validation set. This method is easy to compute. However, it has a high variance because it randomly assigns instances to two datasets and irrespective of the size. Thus, the data points in the validation set are not known and obtain various results for different sets. However, it might produce false results as everything can be executed in a single run. Generally, this method applies to large datasets and may not give good results for small datasets. With small datasets, since there is not sufficient data, the validation process may suffer from underfitting. Also, having a small training dataset will lose important features in the dataset, which increases the error induced by bias.

- K-Fold Cross Validation

A resampling technique called K-fold cross-validation is used to assess learning models using a small dataset. The process has a parameter, k, that designates how many groups a specific data sample should be divided into. K-fold cross-validation uses a sample of data for model training and leaves sufficient data for validation, such that it repeats the hold-out method k times. Generally, a K value of 5 or 10 is used based on experimental evidence; however, it can take any value. Here, k is a reference to the model, such that $k = 5$ becomes 5-fold cross-validation.

This method utilizes a controlled set of instances to measure the performance in making predictions of unseen data. This is widely used due to its simplicity and usually yields a less skewed or overly optimistic estimation of the model skill than other approaches, including a straightforward train:test split ratio. The majority of the data are utilized for fitting, which greatly lowers bias, and the majority of the data are also used in validation sets, which significantly lowers variance. The efficiency of this strategy is also increased by switching the training and test datasets.

Tasks associated with K-fold cross validation.

1. Consider random data instances.
2. Divide the data instances into k subsets of approximately equal size
3. For each subset do the following.
   3.1. Get one of the k subsets as a hold-out or test, validation dataset.
   3.2. Get the other remaining $(k-1)$ subsets to train the model.
   3.3. Use a model to train the data, then assess it against the test data.
   3.4. Record the performance result and remove the model.
4. Utilizing a sample rate of model performance, review the model's expertise.

In order to determine the model's overall efficacy, the error estimation is averaged across all k trials. As a result, each data point will appear exactly once in the validation set and $k-1$ times in the training set. Accordingly, as shown in Figure 7.4, the training and validation datasets are different in each iteration. Since the dataset is divided into a training and a validation dataset iteratively, the following advantages can be achieved.

- Avoids overfitting.
- Rotation estimation, or out-of-sample testing.
- Assess the generalizability of the result to an independent data set.

Stratified K-Fold Cross-Validation

1st split	Fold 1*	Fold 2	Fold 3	Fold 4	Fold 5	Iteration 1
2nd split	Fold 1	Fold 2*	Fold 3	Fold 4	Fold 5	Iteration 2
3rd split	Fold 1	Fold 2	Fold 3*	Fold 4	Fold 5	Iteration 3
4th split	Fold 1	Fold 2	Fold 3	Fold 4*	Fold 5	Iteration 4
5th split	Fold 1	Fold 2	Fold 3	Fold 4	Fold 5*	Iteration 5

* Denotes the test set

FIGURE 7.4   Data fold generation in k-fold cross-validation.

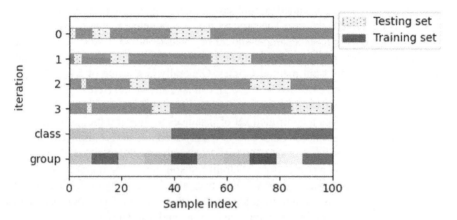

FIGURE 7.5   Stratified k-fold validation representation.

Stratified K-fold uses almost the same percentage of instances from each target class as the total set and addresses the imbalance data problem as shown in Figure 7.5. Therefore, it is used when the distribution of the target variable is not consistent, which would require binning the target variable.

- Leave-P-Out Cross-Validation

Leave-P-out cross-validation drops p number of instances from the training data from a total of n number of data instances. This resulting in n-p instances being utilized for training and the remaining p instances being used as the validation set. This is carried out for each combination obtained by dividing the total number of instances. The entire model efficacy is determined by taking the average error across all trials. This method is extensive since every potential combination of the model must be trained and validated, and for fairly high p, it may become computationally impractical. In practice, p is assigned to 1 in most of the applications and it is defined as cross-validation with leave-one-out. Since the number of viable combinations is equal to the number of data instances in the original sample size, this method is typically selected as it requires less computational work.

7.2.9 KAPPA SCORE

As a performance indicator for a deep learning model, Cohen's kappa is a numerical measure of agreement between two raters. It evaluates the level of agreement between the classification model and the observer in the real-world, both perfectly and randomly as given in (7.9). Kappa score lies between -1 and +1. A value of zero and one implies a random and complete agreement among raters, respectively. When the score is negative, there is less agreement among raters. This gives convincing performance metrics of the model, when we are having an imbalanced dataset.

$$\text{Kappa score} = \frac{\text{Total Accuracy} - \text{Random Accuracy}}{1 - \text{Random Accuracy}} \tag{7.9}$$

where

$$\text{Total accuracy} = \frac{\text{TP} + \text{TN}}{\text{TP} + \text{FP} + \text{FN} + \text{TN}} \text{ and}$$

$$\text{Random accuracy} = \frac{(\text{TN} + \text{FP})(\text{TN} + \text{FN}) + (\text{FN} + \text{TP})(\text{FP} + \text{TP})}{(\text{TP} + \text{TN} + \text{FP} + \text{FN})^2}$$

7.2.10 GRAD-CAM HEAT MAP

Gradient-weighted class activation mapping (Grad-CAM) is an explainable and interpretable technique for CNN models. It increases the transparency and trustworthiness of the model results by highlighting the important regions of an image using the gradients of the target. Grad-CAM generates localization maps of the regions that have more impact for the final decision in an image classification task. Thus, supporting the explainability of the model's decision. Figure 7.6 shows an example of images based on Grad-CAM explanations, where (a) denotes the input image, (b) shows the class activation map, (c) represents the Grad-CAM output and (d) shows the output of Grad-CAM++.

A Grad-CAM representation can be obtained by taking the feature map of the final layer and the weight of each channel in that feature with the gradient of the class classification for a given channel. This shows the activation of the input image for different channels respective to the class showing the region of an image that

(a)	(b)	(c)	(d)

FIGURE 7.6   Grad-CAM based explanations.

impacted to predict the output class. The final score depends primarily on the data in the locations where this gradient is large. This does not need retraining or modifying the existing architecture. Although, Grad-CAM is class discriminative and capable of localizing important regions of an image it cannot emphasize fine-grained information, such as pixel space gradient visualizations. This can be addressed by guided backpropagation, which suppresses the negative gradients when backpropagation through the ReLU layer. Thus, it will capture pixels identified by the nodes.

7.2.11 METRICS FOR IMBALANCED DATASETS

Data imbalance affect to degrade model performance. When a minority class has a low quantity of data in comparison to the other classes, there is an imbalance with the majority class. Some of the techniques that can apply for imbalanced data are listed as follows.

- Youden's index (YI)

Youden's index (YI), also known as Bookmaker informedness (BM) is a frequently used metric of ROC curve. It evaluates a diagnostic marker's efficacy and makes it possible to choose the best threshold value or the cutoff point for the marker. Despite being continuous and positive, certain markers have a peak or positive probability mass at the value of zero. This evaluates how well the algorithm can prevent failure. Youden's index is defined as in (7.10) and a higher score denotes a strong ability to avoid failures. However, it does not alter when the sensitivity and specificity metrics are compared.

$$\text{Youden's index} = \text{sensitivity} - (1 - \text{specificity}) \tag{7.10}$$

- Geometric mean (G–mean)

This is a metric that is used to compare the performance of majority and minority classes. The poor performance of the classifier is indicated through the low geometric mean, which is defined as in (7.11).

$$\text{G-mean} = \sqrt{\frac{\text{TP}}{\text{TP} + \text{FN}} \times \frac{\text{TN}}{\text{TN} + \text{FP}}} = \sqrt{\text{specificity} \times \text{sensitivity}} \tag{7.11}$$

- Likelihood Ratio

Likelihood ratio is used for diagnostic testing and is dependent on both sensitivity and specificity. This ratio is used to determine how a result affects likelihood. All positive findings are not genuine positives, and all negative results are not true negatives in diagnostic testing. Therefore, the likelihood of developing diseases is affected by both positive and bad outcomes. The positive likelihood with a greater value and the negative likelihood with a lower value indicates the superior performance of the positive and negative classes, respectively, and calculate as in (7.12) and (7.13). With balanced and imbalanced datasets, both positive and negative likelihoods are appropriate.

$$\text{Likelihood positive}(\rho+) = \frac{\text{sensitivity}}{1 - \text{specificity}} \qquad (7.12)$$

$$\text{Likelihood negative}(\rho-) = \frac{1 - \text{sensitivity}}{\text{specificity}} \qquad (7.13)$$

- Matthew's Correlation Coefficient (MCC)

The Matthew's correlation coefficient (MCC) is a trustworthy statistical measure. It yields a high score if the forecast successfully predicted the values of TP, FN, TN, and FP, proportionally to the magnitude of both positive and negative items in the dataset, as shown in (7.14). Being a metric to evaluate binary classification task, a high score for MCC can only be achieved if the binary predictor was successful in properly predicting the majority of positive data points and the majority of negative data points.

This is a contingency matrix technique of generating the Pearson product-moment correlation coefficient between actual and predicted values, which is a measure unaffected by the unbalanced datasets problem. It falls between [1, +1], reaching extreme values of -1 and +1, in the case of perfect misclassification and perfect classification, respectively. When a row or column is entirely 0, it is undefinable.

$$\text{MCC} = \frac{\text{TP.TN} - \text{FP.FN}}{\sqrt{(\text{TP}+\text{FP}).(\text{TP}+\text{FN}).(\text{TN}+\text{FP}).(\text{TN}+\text{FN})}} \qquad (7.14)$$

7.2.12 METRICS FOR REGRESSION PROBLEMS

- Mean Absolute Error (MAE)

Mean absolute error (MAE) calculates the average of the disparity between the real and expected outputs. It calculates the accuracy of the forecasts compared to the actual result and averages the absolute errors as in (7.15). The lower metric value for MAE is better.

$$\text{MAE} = \frac{1}{n}\sum_{i=1}^{n}|y - \hat{y}| \qquad (7.15)$$

where n is the total number of data points, y is the actual output, and $\hat{y}$ is the predicted output.

- Mean Squared Error (MSE)

Mean squared error (MSE) calculates the average of the squared variance among the initial and expected outputs as in (7.16). The MSE is a positive value, and the values closer to zero (lowest) are better.

$$\text{MSE} = \frac{1}{n} \sum_{i=1}^{n} (y - \hat{y})^2 \tag{7.16}$$

where n, y, and $\hat{y}$ denote the total number of data points, actual and predicted output, respectively.

- Root Mean Square Error (RMSE)

Root mean square error (RMSE) calculates the square root of mean squared error as given in (7.17). The disparities between the model predicted values and the actual values are quantified by this term, which is also known as the root mean squared deviation. While MAE assigns all errors the same weight, the RMSE penalizes uncertainty by giving greater absolute value errors more weight than smaller absolute value errors. The RMSE is never less than the MAE since all metrics are evaluated.

$$\text{RMSE} = \sqrt{\frac{1}{n} \sum_{i=1}^{n} (y - \hat{y})^2} \tag{7.17}$$

Where n, y, and $\hat{y}$ denote the total number of data points, actual and predicted output, respectively.

- Logarithmic Loss (Log Loss)

Logarithmic loss indicates the closeness of the prediction probability to the actual/ corresponding true value. The maximization of the likelihood is equivalent to minimizing the MSE. It accounts for the uncertainty in model projections by penalizing the incorrect predictions. This performs well for multi-class classification and lower values give better performances as shown in Figure 7.7. Here, the model sets a probability for every class. This supports comparing the performance of the two models. Log loss exists between $[0, \infty)$ and has no upper bound. Higher accuracy is indicated by a log loss that is closer to zero, whereas a log loss that is further from zero suggests lower accuracy. In general, the classifier is more accurate when log loss is minimized. This can be defined as in (7.18) or (7.19).

$$\text{Log loss} = -\frac{1}{N} \sum_{i=1}^{n} \left(\log(p_i) * y_i \right) + \left((1 - y_i) * \log(1 - p_i) \right) \tag{7.18}$$

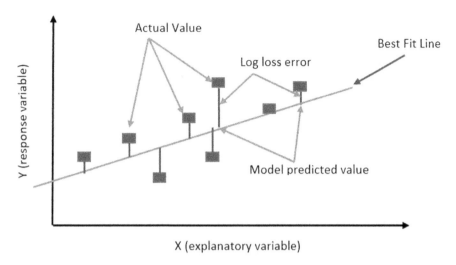

FIGURE 7.7   Representation of log loss error.

where, N is the number of observations, and p is the predicted probabilities that observation i is 1, y is the true label (0 or 1) for observation I, in binary classification.

This can be also represented by

$$\text{Log loss} = -\frac{1}{N}\sum_{i=1}^{N}\sum_{j=1}^{M}\left(\log\left(p_{ij}\right)*y_{ij}\right)$$
(7.19)

where N samples belong to M classes,
y_{ij} indicates whether sample i belongs to class j or not
p_{ij} indicates the probability of sample i belonging to class j

- R Squared Error

R squared or the coefficient of determination shows how closely the model's forecasts match the actual data. This is calculated as in (7.20). R squared has an understandable value and is not affected by the actual output, and does not provide any insight into the prediction error.

$$\text{R squared error} = 1 - \frac{\sum_{i=1}^{n}\left(y_i - \hat{y}_i\right)^2}{\sum_{i=1}^{n}\left(y_i - \bar{y}\right)^2}$$
(7.20)

where y, $\hat{y}$, and $\bar{y}$ denote the actual, predicted, and the mean of the actual values.

TABLE 7.1
Summary of Performance Metrics

Technique	Description
Accuracy	It is mostly used when all the classes are equally important, and the problem is balanced
Precision	Precision becomes crucial for extremely unbalanced datasets with significantly more negatives than positives because they are easy to get high accuracy for (always predicting negative would result in high accuracy).
Recall	When the cost of a false negative is high, should use recall.
F1-measure	In order to find a balance between precision and recall in an unequal class distribution, F1-score can be used to utilize a task with many TNs.
Mean absolute error (MAE)	When performance is evaluated using continuous data, MAE is typically used. It produces a linear number that evenly weights the individual differences and averages them. The model performs better the lower the value.
Mean squared error (MSE)	MSE is used to assess regression models that are used to predict dependent variables that are numeric, while modeling independent variables and dependent or target variables.
ROC	Do not use it when data is heavily imbalanced.
AUC	Use when caring equally about positive and negative classes.
PR-AUC	When data is imbalanced and when care more about positive than negative class.
Hold-out method	One train–test split is all that the hold out approach requires. As a result, the score of the hold-out approach depends on how the data is divided into the train and test sets. When a dataset is really huge, this method is useful.
k-cross-validation	k-cross validation is a resampling method used to assess a model when we have less data and to examine the efficacy of the learning models.
Leave-one-out cross-validation	When the dataset is small or when an accurate assessment of model performance is more important than the computing cost of the method, the leave-one-out cross-validation procedure is appropriate.
Confusion matrix	When the output of the classifier has two or more classes, the confusion matrix is used. Important predictive analytics like accuracy, recall, precision, and specificity are visualized using the associated TP, TN, FP, FN values in the confusion matrix.
Sensitivity	The percentage of accurately categorized positive instances is the sensitivity. It assesses how well a classification performs when the data is unbalanced.
Specificity	The percentage of the negative instances that are accurately categorized is known as the specificity. This can be used with imbalanced data.
Kappa score	The kappa score calculates the prediction's agreement with the true class
Logarithmic loss	Log loss, can compare the performance of two models and then decide which model performs better.

7.2.13 SUMMARY OF PERFORMANCE METRICS

The effectiveness of the performance metrics depends on the data types, application domain, and other constraints. Table 7.1 represents a summary of performance metrics.

REVIEW QUESTIONS

1. What is the ROC curve and what does it represent?
2. When there are many false positives or false negatives, how does it impact the model?
3. What is overfitting and what are the methods that can be used to prevent it?
4. When the values of k becomes larger in K-fold cross-validation, can it result in overfitting or underfitting? Explain your answer.
5. Explain the confusion matrix concerning machine learning algorithms.
6. What is the best way to measure performance improvement on imbalanced datasets?
7. What are the pain points of Cohen's Kappa?

Appendix – Frequently Asked Questions

1. What are hyperparameters in a deep learning model? What are their importances?

   Learning rate: generally, training should be done starting with a small learning rate and increasing the learning rate exponentially for every batch and plotting the loss against the learning rate. The point of the fastest decrease in the loss can be determined by observing the graph.

   The number of epochs: this indicates the number of times the weights are changed in the learning model. Therefore, it is better to train with a large number of epochs such as 50, 100, 150, and 200 to get better convergence. In situations where the number of epochs cannot be determined based on a generalized manner, it is better to use early stopping methods where each network is trained with an initial number of epochs and observe whether training MSE is stuck in a minimum, if not the number of echoes can be increased.

   Batch size: every model responds differently to different batch sizes. For instance, in GPU acceleration, training can physically become faster, with the increase of the batch size until the saturated GPU load. Decreasing batch size can also affect performance, either positively or negatively, if the network has BatchNorm layers.

   Optimizers: techniques such as Adam and SGD can bring better results if combined with a good learning rate and annealing schedule, which aims to manage its value during the training. Hence, it would be better to try different optimizers such as Adam, RMSProp, and SDG, and observe the results.

   Loss function: the widely used loss function is cross-entropy loss or log loss. It decreases as the predicted probability converges to the actual label. It measures the performance of a classification model whose predicted output is a probability value between 0 and 1.

   Activation function: generally, ReLU is a better choice for hidden layers. When it results in the dying ReLU problem then its modifications like leaky ReLU, ELU, and SELU can be used. For binary classification problems, sigmoid is the right choice and for multiclass classification, Softmax is the right choice. However, functions, such as sigmoid and tanh, tend to have vanishing gradient problems.

   Dropout rate: generally, we divide the number of nodes in the layer before dropout by the proposed dropout rate and use that as the number of nodes in the new network that uses dropout. For example, a network with 100 nodes and a proposed dropout rate of 0.5 will require 200 nodes (100/0.5) when using dropout.

2. What is meant by the convergence curve?
   A plot of model learning performance across time or experience is known as a learning curve or convergence curve. For algorithms that gain knowledge incrementally from a training dataset, learning curves are an often-used diagnostic tool in machine learning.

3. What is a gradient explosion?
   Exploding gradient problem acquires high error gradients and results in weight updates with high values during training. As a result, your model can become unstable and unable to benefit from your training set of data.

4. What are the activation functions and their uses?
   Artificial neurons receive inputs and weights from which they calculate the weighted sum of the input, and then it is given to an activation function that converts it into the output. Thus, an activation function is used to map the input to the output, by helping the neural network to learn complex relationships and patterns in data. For instance, ReLU is used for hidden layers, sigmoid is used for binary classification and Softmax is used for multi-class classification.

5. What are the regularization techniques and their uses?
   To prevent overfitting with a twofold stopping condition, this is utilized as an early stopping mechanism. The validation set must not have improved after n epochs, hence the maximum number of epochs is used.

6. What is Xavier initialization?
   This method initializes the weights using a constant variance of the activations across each layer. This prevents the gradient from exploding or vanishing, by allowing signals to reach deep into the network. If the weights in a network start too small, then the signal shrinks as it passes through each layer until it becomes very low to be useful.

7. What if all the weights are initialized with the same value?
   The same signal will be sent to all concealed layers. For instance, each unit receives a signal equal to the sum of inputs if all weights are initialized to 1.

8. What is the disadvantage of using zero initialization for weight initialization?
   The derivatives are constant for each weight if all the weights are initially set to zero. This will result in network failure or symmetry break, because it learns the identical features in each iteration. Any constant initialization will produce low results.

9. How to decide the learning rate.
   A high learning rate may converge the learning model to a suboptimal solution within less time. This can cause undesirable divergent behavior in the loss function. A small learning rate can cause the process to get slow or stuck as the weights update at a low rate. In a model training with constrained time

duration, it would be better to tune the learning rate as it is the most important hyperparameter.

10. What is meant by overfitting?
A statistical modeling error called overfitting happens when a function is tightly matched to a small number of data points. Thus, attempting to make the model conform too closely to slightly inaccurate data can infect the model with substantial errors and reduce its predictive power.

11. How to prevent overfitting or underfitting.
Use approaches such as cross-validation, training with a large amount of data, augmenting data instances, reducing complexity, and early stopping.

12. What is batch normalization?
Batch normalization standardizes the inputs to a network. It directly applies to the activations of a prior layer or the inputs. This is used to accelerate the training process. It reduces the generalization error by decreasing the epochs or providing regularizations.

13. How to select a DL model for a given problem.
The selection of the model is based on different factors, such as the modality of the dataset (image, video), the dimensionality of the data (text, images 2D, for fMRI 4D, MRI 3D, etc.), feature types (aggregations, linear, time series). The potential cluster of architectures can be trained and compared to the results to select the best out of the selection. Some of the features of CNN-based models can be listed as follows.

ResNet: incorporates skip connections between layers and utilizes batch normalization to normalize the input of activation functions. These architectures make it possible to efficiently train an exceptionally deep neural network.

VGG & AlexNet: requires large memory usage and computational power due to the associated large parameter space with many hidden layers.

Inception family models (GoogLeNet, Inception, Xception): provide efficient models.

GoogLeNet: good accuracy with the low memory requirement.

Xception: the accuracy is similar to ResNet. Complicated to apply modifications.

MobileNet: use to execute on mobile devices. This uses few parameters and requires less memory and is computationally simpler. MobileNet net gives a good trade-off between model size and accuracy.

14. What are the pretrained datasets?
ImageNet
MNIST
PaHaW dataset
AlexNet-ImageNet
CIFAR-10

15. What is the reason for using pretrained weights?

    Pretrained weights can be used to overcome the limitation of a small dataset.

    However, in this process, we need to train only the later set of layers, by freezing the remaining layers.

16. What are the models supported by the pretrained weights?

    Several CNN-based models are supported, such as ResNet50, GoogLeNet, VGG19, VGG16, InceptionNet v3, and ResNet18.

17. What are the general processing steps with pretrained models?

    Usually using transfer learning will give better results and faster convergence.

    Load pretrained weights in the constructor.

    Only make the last few convolutional layers trainable.

    Use SGD or another optimizer and set the learning rate low.

    Set the correct activation function at the output layer.

    Increase epochs.

    Fine-tuning the pretrained model, and freezing the weights of the first few layers.

18. What is transfer learning and what are its uses?

    Transfer learning aims to obtain better performance. When a model is trained with a dataset, the associated weights can be initialized randomly or by using Xavier. In transfer learning, the weights can be initialized to comply with the previous task. For instance, ImageNet, and gradient descent can be used to find the best weights. Since the initial value is already meaningful, transfer learning trains the model faster and with fewer data points.

    Fine-tune process consists of starting with the pretrained weights, freezing them, and running gradient descent only on the added new neurons, until convergence. Then unfreeze the pretrained weights and run gradient descent until convergence for the whole network. Generally, a small learning rate can be used for the second gradient descent, allowing pretrained weights to change.

19. How to increase the model accuracy.

    The model accuracy relies on aspects such as the learning model, size of the dataset and image augmentations, and tuning of hyperparameters. Some of the possible tasks can be listed as follows.

    Select a good dataset with balanced and not biased data points.

    Obtain the model convergence by tuning the learning rate starting with a lower rate and tuning other hyperparameters.

    Use different weight initialization for the weights.

    Train with more epochs.

20. Although the testing accuracy is high, what is the reason for getting inaccurate classification when testing one image at a time?

    When the model gets overfit for a set of test data, the results in that sample become high. The following methods can be applied to mitigate the model overfitting.

The training and testing datasets should be balanced dataset to represent all potential classes with nearly equal distribution.

Generalize the accuracy by dividing the entire data instances into training, testing, and validation in the ratio.

Use cross-validation techniques.

21. What is the reason for getting high testing errors than training errors?
This indicates the model is overfitting. The regularization technique such as early stopping with dropout and cross-validation can be used to reduce overfitting. The other possible techniques are data augmentation, unimportant feature removal, regularization, and model ensembling.

22. Distinguish the training, testing, and validation datasets.
The training dataset is used to fit the model training.

The validation dataset is utilized for a fair assessment of a model that performs well on the training dataset while tuning model hyperparameters. Different samples are included in validation datasets to assess trained ML models. The model can still be adjusted and managed at this point.

A test dataset is a distinct sample that offers a final, unbiased assessment of a model's fit. The inputs are comparable to the stages before it, however, they are not the same data.

23. What are the best-split ratios:
Common split ratios for training, testing, and validation are,
70% training, 15% validation, 15% testing
80% training, 10% validation, 10% testing
60% training, 20% validation, 20% testing

24. What are the platforms that we can use to implement DL models?
This depends based on the model size, dataset size, and batch size of the training procedure.

Jupyter notebook: difficult to work with models that require high computational power,

Tensor Hub

Google Colab: have memory limitations in the longer run. For example, it may not work well with models such as mask-RCNN and BERT pretraining tasks that need more memory and training time.

25. What are the frameworks that we can use to implement DL models? How do we select a framework to use? What are the conditions and features we should consider?
Keras is used when developers seek a plug-and-play framework that enables them to design, train, and evaluate their models rapidly. It offers more deployment options and easier model export. However, PyTorch is faster than Keras and has better debugging capabilities.

Pytorch is a better framework for developing DL models for research due to its compatibility with parallel processing API. In contrast, TensorFlow provides an easy pipeline with TensorFlow Lite to develop lightweight models for edge inference.

However, edge platforms, such as NCS2, are also compatible with many DL frameworks including PyTorch and TensorFlow.

26. What are the useful libraries that we can use and what are the problems we can solve using those libraries?
NumPy: A Python library for working with arrays, matrices, the Fourier transform, and the area of linear algebra.

SciPy: Python library that is open-source and used for technical and scientific computing. It contains modules for optimization, linear algebra, integration, interpolation, special functions, FFT, signal and image processing, ODE solvers and other tasks common in science and engineering.

Sklearn: Opensource machine learning library for the Python programming language. It features various classification, regression and clustering algorithms, including support vector machines, random forests, gradient boosting, k-means and DBSCAN, and is designed to interoperate with the Python numerical and scientific libraries.

Pandas: a Python-based open-source data analysis and manipulation tool that is quick, strong, adaptable, and simple to use.

Keras: on top of the machine learning platforms, Keras is an open-source, Python-based deep learning API.

Tensor Hub: provides a repository of pretrained machine learning models as off-the-shelf models to be used in machine learning tasks.

XGBoost: provides an optimized machine learning algorithm under the gradient boosting framework.

27. How to increase the inference speed of the model
One of the more important considerations when implementing a deep network in a production setting is network latency. Few options to obtain fast inference time: extract every last drop out of our model, use a field-programmable gate array (FPGA) to provide high throughput, frame per second (FPS) and lower latency compared to GPUs.

Model pruning: this discards insignificant weights to get a smaller and faster learning model. In neural networks, both convolution and fully connected layers can be pruned. But convolution layers are more sensitive to pruning than fully connected layers. Thus, a threshold can be chosen according to the sensitivity of each layer.

Use a field-programmable gate array (FPGA) to provide high throughput, frame per second (FPS) and lower latency compared to GPUs.

28. What is the difference between prediction and inference?
The process of applying a trained learning algorithm to produce a prediction is known as inference. It encourages coming to a decision that has been justified by facts and evidence.

A prediction is a firm assertion about what will happen in the future.

29. How to decide the trade-off between model throughput and accuracy.
In machine learning, throughput is a metric used to assess how well various models perform in a given application. The quantity of data units processed in a certain amount of time is referred to as throughput.

The ratio of the total number of predictions produced to the number of classifications a model properly predicts is known as model accuracy. It is a method of rating a model's effectiveness.

30. How to select an evaluation technique?
Every machine learning problem can be categorized into two categories of classification and regression. Based on the category the evaluation metric can be different to be applied.

Classification metrics: Confusion matrix representing, Accuracy, precision, and recall, F-score.

Regression metrics: Explained variance, Mean squared error, R2 coefficient.

Common metrics: Learning curves, Validation curves.

References

1. Abdel-Jaber, H., Devassy, D., Al Salam, A., Hidaytallah, L., EL-Amir, M., 2022. A review of deep learning algorithms and their applications in healthcare. Algorithms 15, 71. doi: 10.3390/a15020071.

2. Abeysinghe, C., Perera, I., Meedeniya, D., 2021. Capsule networks for character recognition in low resource languages, in: Malarvel, M., Nayak, S.R., Pattnaik, P.K., Panda, S.N. (Eds.), Machine vision inspection systems, Volume 2: Machine Learning-Based Approaches. John Wiley and Sons. chapter 2, pp. 23–46. doi: 10.1002/9781119786122. ch2.

3. Agarwal, N., Sondhi, A., Chopra, K., Singh, G., 2021. Transfer learning: Survey and classification. Smart innovations in communication and computational sciences, 145–155. doi: 10.1007/978-981-15-5345-5 13.

4. Agarwal, V., Lohani, M., Bist, A.S., Harahap, E.P., Khoirunisa, A., 2022. Analysis of deep learning techniques for chest x-ray classification in context of covid-19. ADI Journal on Recent Innovation 3, 208–216. doi: 10.34306/ajri.v3i2.659.

5. Al Husaini, M.A.S., Habaebi, M.H., Gunawan, T.S., Islam, M.R., Elsheikh, E.A., Suliman, F., 2022. Thermal-based early breast cancer detection using inception v3, inception v4 and modified inception mv4. Neural Computing and Applications 34, 333–348. doi: 10.1007/s00521-021-06372-1.

6. Alzubaidi, L., Zhang, J., Humaidi, A.J., Al-Dujaili, A., Duan, Y., Al-Shamma, O., Santamaría, J., Fadhel, M.A., Al-Amidie, M., Farhan, L., 2021. Review of deep learning: Concepts, CNN architectures, challenges, applications, future directions. Journal of Big Data 8, 1–74. doi: 10.1186/s40537-021-00444-8.

7. Ariyarathne, G., De Silva, S., Dayarathna, S., Meedeniya, D., Jayarathne, S., 2020. ADHD identification using convolutional neural network with seed-based approach for fMRI data, in: Proceedings of 9th International Conference on Software and Computer Applications (ICSCA), pp. 31–35. doi: 10.1145/3384544.3384552.

8. Bandara, M., Jayasundara, R., Ariyarathne, I., Meedeniya, D., Perera, C., 2023. Forest sound classification dataset: FSC22, Sensors, 23, 4:2032, doi: 10.3390/s23042032.

9. Belousov, B., Abdulsamad, H., Klink, P., Parisi, S., Peters, J., 2021. Reinforcement learning algorithms: Analysis and applications. Springer.

10. Bozinovski, S., Fulgosi, A., 1976. The influence of pattern similarity and transfer learning upon training of a base perceptron b2, in: Proc. Symposium Informatica, pp. 121–126. doi: 10.31449/inf.v44i3.2828.

11. Brendan McMahan, H., Moore, E., Ramage, D., Hampson, S., Ag¨uera y Arcas, B., 2016. Communication-efficient learning of deep networks from decentralized data. arXive-prints, arXiv–1602 doi: 10.48550/arXiv.1602.05629.

12. Brownlee, J., 2018. Better deep learning: train faster, reduce overfitting, and make better predictions. Machine Learning Mastery.

13. Chauhan, N.K., Singh, K., 2018. A review on conventional machine learning vs deep learnin g, in: Proc. International conference on computing, power and communication technologies (GUCON), IEEE. pp. 347–352. doi: 10.1109/gucon.2018.8675097.

14. Chitty-Venkata, K.T., Somani, A.K., 2022. Neural architecture search survey: A hardware perspective. ACM Computing Surveys (CSUR), 55(4):78, PP. 1-36. doi: 10.1145/3524500.

15. Chollet, F., 2017. Xception: Deep learning with depth-wise separable convolutions, in: Proc. International Conference on Computer Vision and Pattern Recognition (CVPR), pp. 1800–1807. doi: 10.1109/CVPR.2017.195.

16. Dasanayaka, S., Shantha, V., Silva, S., Ambegoda, T., Meedeniya, D., 2022a. Interpretable machine learning for brain tumor analysis using MRI, in: Proceedings of the 2nd International Conference on Advanced Research in Computing (ICARC), pp. 212–217. doi: 10.1109/ICARC54489.2022.9754131.

17. Dasanayaka, S., Shantha, V., Silva, S., Meedeniya, D., Ambegoda, T., 2022b. Interpretable machine learning for brain tumour analysis using MRI and whole slide images. Software Impacts 13, 100340. doi: 10.1016/j.simpa.2022.100340.

18. De Silva, S., Dayarathna, S., Ariyarathne, G., Meedeniya, D., Jayarathna, S., Michalek, A.M., 2021. Computational decision support system for ADHD identification. International Journal of Automation and Computing (IJAC) 18, 233–255. doi: 10.1007/s11633-020-1252-1.

19. De Silva, S., Dayarathna, S., Meedeniya, D., 2022. Alzheimer's disease diagnosis using functional and structural neuroimaging modalities, in: Wadhera, T. and Kakkar, D. (Ed.), Enabling technology for neurodevelopmental disorders from diagnosis to rehabilitation. Taylor and Francis CRS Press, Routledge. chapter 11, pp. 162–183. doi: 10.4324/9781003165569-11.

20. De Silva, S., Dayarathna, S.U., Ariyarathne, G., Meedeniya, D., Jayarathna, S., 2021b. fMRI feature extraction model for ADHD classification using convolutional neural network. International Journal of E-Health and Medical Communications (IJEHMC) 12, 81–105. doi:10.4018/IJEHMC.2021010106.

21. Demotte, P., Wijegunarathna, K., Meedeniya, D., Perera, I., 2021. Enhanced sentiment extraction architecture for social media content analysis using capsule networks. Multimedia Tools and Applications doi: 10.1007/s11042-021-11471-1.

22. Desai, M., Shah, M., 2021. An anatomization on breast cancer detection and diagnosis employing multi-layer perceptron neural network (MLP) and convolutional neural network (CNN). Clinical eHealth 4, 1–11. doi: 10.1016/j.ceh.2020.11.002.

23. Dong, S., Wang, P., Abbas, K., 2021. A survey on deep learning and its applications. Computer Science Review 40, 100379. doi: 10.1016/j.cosrev.2021.100379.

24. Dosovitskiy, A., Beyer, L., Kolesnikov, A., Weissenborn, D., Zhai, X., Unterthiner, T., Dehghani, M., Minderer, M., Heigold, G., Gelly, S., et al., 2021. An image is worth 16x16 words: Transformers for image recognition at scale. in: Proceedings of the The International Conference on Learning Representations (ICLR) , pp. 1–22.

25. Eelbode, T., Sinonquel, P., Maes, F., Bisschops, R., 2021. Pitfalls in training and validation of deep learning systems. Best Practice & Research Clinical Gastroenterology 52, 101712. doi: 10.1016/j.bpg.2020.101712.

26. Ekman, M., 2021. Learning deep learning: Theory and practice of neural networks, computer vision, NLP, and transformers using TensorFlow. Addison-Wesley Professional.

27. Fernando, C., Kolonne, S., Kumarasinghe, H., Meedeniya, D., 2022. Chest radiographs classification using multi-model deep learning: A comparative study, in: Proceedings of the 2nd International Conference on Advanced Research in Computing (ICARC), pp. 165–170. doi: 10.1109/ICARC54489.2022.9753811.

28. Goodfellow, I., Pouget-Abadie, J., Mirza, M., Xu, B., Warde-Farley, D., Ozair, S., Courville, A., Bengio, Y., 2020. Generative adversarial nets, Communications of the ACM 63, 139–144, doi: 10.1145/342262.

29. G´eron, A., 2018. Neural networks and deep learning. O'Reilly Media, Inc. He, K., Zhang, X., Ren, S., Sun, J., 2016. Deep residual learning for image recognition, in: Proc.

IEEE Conference on Computer Vision and Pattern Recognition (CVPR), pp. 770–778. doi: 10.1109/CVPR.2016.90.

30. He, K., Zhang, X., Ren, S., Sun, J., 2016. Deep residual learning for image recognition, in: Proc. IEEE Conference on Computer Vision and Pattern Recognition (CVPR), pp. 770–778. doi:10.1109/CVPR.2016.90.

31. Herath, L., Meedeniya, D., Marasingha, J., Weerasinghe, V., 2021. Autism spectrum disorder diagnosis support model using inceptionv3, in: Proceedings of International Research Conference on Smart Computing and Systems Engineering (SCSE), pp. 1–7. doi: 10.1109/SCSE53661.2021.9568314.

32. Herath, L., Meedeniya, D., Marasingha, J., Weerasinghe, V., 2022. Optimize transfer learning for autism spectrum disorder classification with neuroimaging: A comparative study, in: Proceedings of the 2nd International Conference on Advanced Research in Computing (ICARC), pp. 171–176. doi: 10.1109/ICARC54489.2022.9753949.

33. Howard, A.G., Zhu, M., Chen, B., Kalenichenko, D., Wang, W., Weyand, T., Andreetto, M., Adam, H., 2017. Mobilenets: Efficient convolutional neural networks for mobile vision applications. arXiv preprint. doi: 10.48550/arXiv.1704.04861.

34. Huang, G., Liu, Z., Van Der Maaten, L., Weinberger, K.Q., 2017. Densely connected convolutional networks, in: Proc. International Conference on Computer Vision and Pattern Recognition (CVPR), pp. 2261–2269. doi: 10.1109/CVPR.2017.243.

35. Hutter, F., Kotthoff, L., Vanschoren, J., 2019. Automated machine learning: methods, systems, challenges. Springer Nature. doi: 10.1007/978-3-030-05318-5.

36. Iandola, F.N., Han, S., Moskewicz, M.W., Ashraf, K., Dally, W.J., Keutzer, K., 2016. Squeezenet: Alexnet-level accuracy with 50x fewer parameters and! 0.5 mb model size. arXiv preprint, 1–13. doi: 10.48550/arXiv.1602.07360.

37. Joseph, A.D., Nelson, B., Rubinstein, B.I.P., Tygar, J.D., 2019. Adversarial machine learning. Cambridge University Press. doi: 10.1017/9781107338548.

38. Kang, M., Ko, E., Mersha, T.B., 2022. A roadmap for multi-omics data integration using deep learning. Briefings in Bioinformatics 23, bbab454. doi: 10.1093/bib/bbab454.

39. Kapadnis, S., Tiwari, N., Chawla, M., 2022. Developments in capsule network architecture: A review. Intelligent Data Engineering and Analytics 266, 81–90. doi: 10.1007/978-981-16-6624-7_9.

40. Ketkar, N., Santana, E., 2017. Deep learning with Python. volume 1. Springer. doi: 10.1007/978-1-4842-2766-4.

41. Kumar, S., Kaur, P., Gosain, A., 2022. A comprehensive survey on ensemble methods, in: Proc. International conference for Convergence in Technology (I2CT), pp. 1–7, doi: 10.1109/I2CT54291.2022.9825269.

42. Kumarasinghe, H., Kolonne, S., Fernando, C., Meedeniya, D., 2022. U-net based chest x-ray segmentation with ensemble classification for COVID-19 and pneumonia. International Journal of Online and Biomedical Engineering (iJOE) 18, 161–174. doi: 10.3991/ijoe.v18i07.30807.

43. Ladosz, P., Weng, L., Kim, M., Oh, H., 2022. Exploration in deep reinforcement learning: A survey. Information Fusion 85, 1–22. doi: 10.1016/j.inffus.2022.03.003.

44. Laxmisagar, H., Hanumantharaju, M., 2022. Detection of breast cancer with lightweight deep neural networks for histology image classification. Critical Reviews™ in Biomedical Engineering 50, 1–19. doi: 10.1615/CritRevBiomedEng.2022043417.

45. Liu, X., Faes, L., Kale, A.U., Wagner, S.K., Fu, D.J., Bruynseels, A., Mahendiran, T., Moraes, G., Shamdas, M., Kern, C., et al., 2019. A comparison of deep learning performance against health-care professionals in detecting diseases from medical imaging: a systematic review and meta-analysis. The Lancet Digital Health 1, e271–e297. doi: 10.1016/s2589-7500(19)30123-2.

46. Liu, Y., Sun, P., Wergeles, N., Shang, Y., 2021. A survey and performance evaluation of deep learning methods for small object detection. Expert Systems with Applications 172, 114602. doi: 10.1016/j.eswa.2021.114602.

47. Ludwig, H., Baracaldo, N., 2022. Federated learning: A comprehensive overview of methods and applications. Springer Cham. doi: 10.1007/978-3-030-96896-0.

48. Mahakalanda, I., Demotte, P., Perera, I., Meedeniya, D., Wijesuriya, W., Rodrigo, L., 2022. Chapter 7–deep learning-based prediction for stand age and land utilization of rubber plantation, in: Khan, M.A., Khan, R., Ansari, M.A. (Eds.), Application of Machine Learning in Agriculture. Elsevier Academic Press, pp. 131–156. doi: 10.1016/B978-0-323-90550-3.00008-4.

49. Mandal, M., Vipparthi, S.K., 2021. An empirical review of deep learning frameworks for change detection: Model design, experimental frameworks, challenges and research needs. IEEE Transactions on Intelligent Transportation Systems 23, 6101–6122. doi: 10.1109/tits.2021.3077883.

50. Mathew, A., Amudha, P., Sivakumari, S., 2020. Deep learning techniques: an overview, in: Proc. International conference on advanced machine learning technologies and applications, pp. 599–608. doi: 10.1007/978-981-15-3383-954.

51. Meedeniya, D., Kumarasinghe, H., Kolonne, S., Fernando, C., De la Torre D´ıez, I., Marques, G., 2022a. Chest x-ray analysis empowered with deep learning: A systematic review. Applied Soft Computing, 109319. doi: 10.1016/j.asoc.2022.109319.

52. Meedeniya, D., Mahakalanda, I., Lenadora, D., Perera, I., Hewawalpita, S., Abeysinghe, C., Nayak, S., 2022b. Chapter 13–Prediction of paddy cultivation using deep learning on land cover variation for sustainable agriculture, in: Poonia, R.C., Singh, V., Nayak, S.R. (Eds.), Deep learning for sustainable agriculture. Elsevier Academic Press. pp. 325–355. doi: 10.1016/B978-0-323-85214-2.00009-4.

53. Meedeniya, D., Rubasinghe, I., 2020. A review of supportive computational approaches for neurological disorder identification, in: Wadhera, T., Kakkar, D. (Eds.), Interdisciplinary approaches to altering neurodevelopmental disorder. IGI Global. chapter 16, pp. 271–302. doi: 10.4018/978-1-7998-3069-6.ch016.

54. Nagrath, P., Jain, R., Madan, A., Arora, R., Kataria, P., Hemanth, J., 2021. Ssdmnv2: A real time DNN-based face mask detection system using single shot multibox detector and mobilenetv2. Sustainable Cities and Society 66, 102692. doi: 10.1016/j.scs.2020.102692.

55. Nguyen, D.C., Pham, Q.V., Pathirana, P.N., Ding, M., Seneviratne, A., Lin, Z., Dobre, O., Hwang, W.J., 2022. Federated learning for smart healthcare: A survey. ACM Computing Surveys (CSUR) 55, 1–37. doi: 10.1145/3501296.

56. Nielsen, M.A., 2015. Neural networks and deep learning. Volume 25. Determination press San Francisco, USA.

57. Opitz, D., Maclin, R., 1999. Popular ensemble methods: An empirical study. Journal of Artificial Intelligence Research 11, 169–198. doi: 10.1613/jair.614.

58. Padmasiri, H., Madurawe, R., Abeysinghe, C., Meedeniya, D., 2020. Automated vehicle parking occupancy detection in real-time, in: Proceedings of 2020 Moratuwa Engineering Research Conference (MERCon), pp. 1–6. doi: 10.1109/MERCon50084.2020.9185199.

59. Padmasiri, H., Shashirangana, J., Meedeniya, D., Rana, O., Perera, C., 2022. Automated license plate recognition for resource-constrained environments. Sensors 22. 1434. doi: 10.3390/s22041434.

60. Pathirana, P., Senarath, S., Meedeniya, D., Jayarathna, S., 2022a. Eye gaze estimation: A survey on deep learning-based approaches. Expert Systems with Applications 19, 1–16. doi: 10.1016/j.eswa.2022.116894.

61. Pathirana, P., Senarath, S., Meedeniya, D., Jayarathna, S., 2022b. Single-user 2D gaze estimation in retail environment using deep learning, in: Proc. of the 2nd International Conference on Advanced Research in Computing (ICARC), pp. 206–211. doi: 10.1109/ ICARC54489.2022.9754167.

62. Qian, C., Zhu, J., Shen, Y., Jiang, Q., Zhang, Q., 2022. Deep transfer learning in mechanical intelligent fault diagnosis: Application and challenge. Neural Processing Letters 54, 2509–2531. doi: 10.1007/s11063-021-10719-z.

63. Ravichandiran, S., 2019a. Hands-on deep learning algorithms with Python: Master deep learning algorithms with extensive math by implementing them using TensorFlow. Packt Publishing Ltd.

64. Romo-Montiel, E., Menchaca-Mendez, R., Rivero-Angeles, M.E., Menchaca-Mendez, R., 2022. Improving communication protocols in smart cities with transformers. ICT Express 1, 50–55. doi: 10.1016/j.icte.2022.02.006.

65. Ronneberger, O., Fischer, P., Brox, T., 2015. U-net: Convolutional networks for biomedical image segmentation, in: Proc. International Conference on Medical image computing and computer-assisted intervention, pp. 234–241. doi: 10.1007/ 978-3-319-24574-428.

66. Rubasinghe, I., Meedeniya, D., 2019. Ultrasound nerve segmentation using deep probabilistic programming. Journal of ICT Research and Applications 13, 241–256. doi: 10.5614/itbj.ict.res.appl.2019.13.3.5.

67. Rubasinghe, I., Meedeniya, D., 2020. Automated neuroscience decision support framework, in: Agarwal, B., Balas, V., Jain, L., Poonia, R., Manisha (Eds.), Deep learning techniques for biomedical and health informatics. Elsevier. chapter 13, pp. 305–326. doi: 10.1016/B978-0-12-819061-6.00013-6.

68. Sabour, S., Frosst, N., Hinton, G.E., 2017. Dynamic routing between capsules, in: Proc. International Conference on Neural Information Processing Systems (NIPS), pp. 3859– 3869. doi: 10.48550/arXiv.1710.09829.

69. Sandler, M., Howard, A., Zhu, M., Zhmoginov, A., Chen, L., 2018. Mobilenetv2: Inverted residuals and linear bottlenecks, in: Proc. International Conference on Computer Vision and Pattern Recognition (CVPR), pp. 4510–4520. doi: 10.1109/CVPR.2018.00474.

70. Senarath, S., Pathirana, P., Meedeniya, D., Jayarathna, S., 2022a. Customer gaze estimation in retail using deep learning. IEEE Access 10, 64904–64919. doi: 10.1109/ ACCESS.2022.3183357.

71. Senarath, S., Pathirana, P., Meedeniya, D., Jayarathna, S., 2022b. Retail gaze: A dataset for gaze estimation in retail environments, in: Proceedings of the 3rd International Conference on Decision Aid Sciences and Applications (DASA), pp. 1040–1044. doi: 10.1109/DASA54658.2022.9765224.

72. Sewak, M., Karim, M.R., Pujari, P., 2018. Practical convolutional neural networks: Implement advanced deep learning models using Python. Packt Publishing Ltd.

73. Shafiq, M., Gu, Z., 2022. Deep residual learning for image recognition: A survey. Applied Sciences 12, 8972. doi: 10.3390/app12188972.

74. Shashirangana, J., Padmasiri, H., Meedeniya, D., Perera, C., 2021a. Automated license plate recognition: A survey on methods and techniques. IEEE Access 9, 11203–11225. doi: 10.1109/ACCESS.2020.3047929.

75. Shashirangana, J., Padmasiri, H., Meedeniya, D., Perera, C., Nayak, S.R., Nayak, J., Vimal, S., Kadry, S., 2021b. License plate recognition using neural architecture search for edge devices. International Journal of Intelligent Systems (IJIS) 36, 1–38. doi: 10.1002/int.22471.

76. Shrestha, A., Mahmood, A., 2019. Review of deep learning algorithms and architectures. IEEE Access 7, 53040–53065. doi: 10.1109/access.2019.2912200.

77. Shyamalee, T., Meedeniya, D., 2022a. Attention u-net for glaucoma identification using fundus image segmentation, in: Proceedings of the 3rd International Conference on Decision Aid Sciences and Applications (DASA), pp. 6–10. doi: 10.1109/ DASA54658.2022.9765303.

78. Shyamalee, T., Meedeniya, D., 2022b. CNN based fundus images classification for glaucoma identification, in: Proceedings of the 2nd International Conference on Advanced Research in Computing (ICARC), pp. 200–205. doi: 10.1109/ ICARC54489.2022.9754171.

79. Simonyan, K., Zisserman, A., 2014. Very deep convolutional networks for largescale image recognition. arXiv preprint arXiv:1409.1556, arXiv:1409.1556.

80. Szegedy, C., Liu, W., Jia, Y., Sermanet, P., Reed, S., Anguelov, D., Erhan, D., Vanhoucke, V., Rabinovich, A., 2015. Going deeper with convolutions, in: Proc. International Conference on Computer Vision and Pattern Recognition (CVPR), pp. 1–9. doi: 10.1109/ cvpr.2015.7298594.

81. Szegedy, C., Vanhoucke, V., Ioffe, S., Shlens, J., Wojna, Z., 2016. Rethinking the inception architecture for computer vision, in: Proc. International Conference on Computer Vision and Pattern Recognition (CVPR), pp. 2818–2826. doi: 10.1109/ cvpr.2016.308.

82. Tan, M., Le, Q., 2019. Efficientnet: Rethinking model scaling for convolutional neural networks, in: Proc. International Conference on Machine Learning, pp. 6105–6114. doi: 10.48550/arXiv.1905.11946.

83. Thomas, J.J., Karagoz, P., Ahamed, B.B., Vasant, P., 2019. Deep learning techniques and optimization strategies in big data analytics. IGI Global. doi: 10.4018/ 978-1-7998-1192-3.

84. Ugail, H., 2022. Deep learning in visual computing: Explanations and examples. CRC Press.

85. Vasudevan, S.K., Pulari, S.R., Vasudevan, S., 2022. Deep learning: A comprehensive guide. Chapman and Hall/CRC.

86. Vaswani, A., Shazeer, N., Parmar, N., Uszkoreit, J., Jones, L., Gomez, A.N., Kaiser, L., Polosukhin, I., 2017. Attention is all you need. Advances in Neural Information Processing Systems 30, 1–15. doi: 10.48550/arXiv.1706.03762.

87. Wang, H.n., Liu, N., Zhang, Y.y., Feng, D.w., Huang, F., Li, D.s., Zhang, Y.m., 2020. Deep reinforcement learning: A survey. Frontiers of Information Technology & Electronic Engineering 21, 1726–1744. doi: 10.1631/FITEE.1900533.

88. Wijethilake, N., Meedeniya, D., Chitraranjan, C., Perera, I., 2020. Survival prediction and risk estimation of glioma patients using MRNA expressions, in: Proceedings of 20th International Conference on Bioinformatics and Bioengineering (BIBE), pp. 35– 42. doi: 10.1109/BIBE50027.2020.00014.

89. Wijethilake, N., Meedeniya, D., Chitraranjan, C., Perera, I., Islam, M., Ren, H., 2021. Glioma survival analysis empowered with data engineering—a survey. IEEE Access 9, 43168–43191. doi: 10.1109/ACCESS.2021.3065965.

90. Yan, W., 2021. Computational methods for deep learning. Springer.

91. Yang, Q., Zhang, Y., Dai, W., Pan, S.J., 2020. Transfer learning. Cambridge University Press. doi: 10.1017/9781139061773.

92. Yao, X., Wang, X., Karaca, Y., Xie, J., Wang, S., 2020. Glomerulus classification via an improved googlenet. IEEE Access 8, 176916–176923. doi: 10.1109/ access.2020.3026567.

93. You, A., Kim, J.K., Ryu, I.H., Yoo, T.K., 2022. Application of generative adversarial networks (GAN) for ophthalmology image domains: A survey. Eye and Vision 9, 1–19. doi: 10.1186/s40662-022-00277-3.

94. Zhang, C., Ma, Y., 2012. Ensemble machine learning: methods and applications. Springer. doi: 10.1007/978-1-4419-9326-7.

95. Zhang, J., Li, C., Yin, Y., Zhang, J., Grzegorzek, M., 2022a. Applications of artificial neural networks in microorganism image analysis: a comprehensive review from conventional multilayer perceptron to popular convolutional neural network and potential visual transformer. Artificial Intelligence Review 55, 1–58. doi: 10.1007/s10462-022-10192-7.

96. Zhang, T., Gao, L., He, C., Zhang, M., Krishnamachari, B., Avestimehr, A.S., 2022b. Federated learning for the internet of things: Applications, challenges, and opportunities. IEEE Internet of Things Magazine 5, 24–29. doi: 10.1109/iotm.004.2100182.

97. Zhao, W., Alwidian, S., Mahmoud, Q.H., 2022a. Adversarial training methods for deep learning: A systematic review. Algorithms 15, 283. doi: 10.3390/a15080283.

98. Zhou, T., Ye, X., Lu, H., Zheng, X., Qiu, S., Liu, Y., 2022. Dense convolutional network and its application in medical image analysis. BioMed Research International 2022, 2384830. doi: 10.1155/2022/2384830.

99. Zoph, B., Le, Q.V., 2016. Neural architecture search with reinforcement learning. arXiv preprint arXiv:1611.01578, 1—16. doi: 10.48550/arXiv.1611.01578.

100. Zouch, W., Sagga, D., Echtioui, A., Khemakhem, R., Ghorbel, M., Mhiri, C., Hamida, A.B., 2022. Detection of covid-19 from CT and chest X-ray images using deep learning models. Annals of Biomedical Engineering 50, 825–835. doi: 10.1007/s10439-022-02958-5.

Index

Note: Figures are indicated by *italics*. Tables are indicated by **bold**.

A

accuracy 4, 34, 37, 62–4, 69, 71–2, 82, 88, 99–100, 105, 112–15, 119, 123, 132, 136–7, 139–41, 147–51, 153, 157, 159–60, **162**
activation function 16, 18, 20, 23–9, 34, 39–40, 44–5, 54, 56, 59–61, 66, 71, 123
AdaDelta 130–1
adagrad 130–1
ADAM 131–2
adversarial training 140, 143–4
asynchronous computation 13
agent 91–6
aggregation 97, 100, 105–6
ANN 42, 44–5, 56–8, 73, 76
API 13, 145
argmax 107, 138
artificial intelligence 1–3
attacks 140–5
attention 50, 79–82
AUC 147, 152–3, **162**
augmentation 36, 63, 112, 119–20, 137
autoencoder 76–8
axon 16

B

backpropagation 23, 25–6, 38–9, 47–8, 50, 56, 61, 63, 72, 128, 158
bagging 105–7, 114
batch size 22–3, 121, 123
batch-norm layer 50
bias 6–11, 16, 18–20, 22, 24–5, 30, 33–5, 37–8, 44, **57**, 66, 102, 105–7, 125, 155
binning 112, 156
boosting 106–8, 114, 117
bootstrap 105–6, 114
brain cells 16

C

capsule 73–6
centralize 96, 100–1
chain rule 38
classification 11, 21–30, 37–8, 42–5, 48, 55–8, 61, 66, 71–2, 89–90, 108–10, 119, 140–1, 147–61, 165
CNN 42, 48–58, 62, 66–7, 71–3, 81–2, 103
collaborative learning 96–7

complexity 8–11, 13, 33, 35–7, 40, 55, 58, 67, 78, 105, 110, 116, 121, 124
computational load 66
computer vision 48, **57**, 61–2, 72, 86, 89, 102, 130
confusion matrix 148, **162**
convergence 18, 23, 28, 39, 127–8, 133, 136
convolution 43, 48–9, 51–4, 66–71, 82, 103, 122, 135
cross entropy 30, 77, 139
cross validation 36, 105, 115–16, 125, 153–4, 156

D

DARTS 135, 137–9
data partition 100
data poisoning 141–2
decentralize 100–1
decision boundary 42, 60, 70, 103, 110, 143
decision-making 1–2, 11, 16, 91, 94, 141
decoder 74–82, 135
dense layer 48, 50, 54–5, 70, 114
DenseNet 69–70
depthwise convolution 69–70, 139
dimensionality reduction 69, 76, 78, 113
discriminative model 62–3
distributed features **72**
drift 90
dropout 32, 36, 38, 50, 62, 69, 72, 85, 118–19, 124
dual loss 74
dynamic computation 14, 91, 140
dynamic programming 94
dynamic routing 73–5

E

early stopping 35–7, 87, 117–18
elastic net 117
encoder 74–8, 81–3, 90
ensemble 100, 103–10, 114
epoch 25, 32, 38, 87, 98, 101, 114, 123, 125
evaluation 33, 94, 100, 137, 147, 151
evolution 3–4, 114, 137
evolutionary algorithm 114, 137
evasion 142
exploding 23–4, 26, 38–40, 47–8, **56–7**, 67, 133

F

FBNet 135, 138–9
feature clipping 122

feature extraction 37, 42, 48, **56**, 58, 66, 73, 84, 88
feature selection 21, 42, 113, 143
federated learning 96–102
feedforward 23, 38, 58, *60*, 76
feature engineering 5–6, 42, 113
filter 48, 51–3, 66, 69–72, 110
flatten 49–51, 55, 83
F-measure 151, 154
freeze 85–6
fully connected 28, 49–51, 54, 56, 61, 71, 76, 85, 89, 103, 132, 134–5
fusion 50, 103, 109–10

G

GAN 62–4, 112, 119
Gaussian 24, 27, 78
generalization error 10, 33, 103, 105–6
generalize 8, 35, 37, 115–16, 132
geometric mean 158
global features 72, 97–8, 101, 133
GoogLeNet 67–9
Grad-CAM 157–8
gradient descent 24–6, 34, 39, 94, 115, 121, 125, *127*, 30, 137–8

H

hardware agnostic 139
Hebbian principle 66
heterogeneous 96, 100, 102–3
hidden layer 3–4, 20, 22–3, 26, 28, 36, 39, 43, 45, 50, 52, 58–61, 88, 119, 123–4, 134, 165
hidden patterns 12
high computation 3, **72**
hold-out validation 106, 154, 155, **162**
Huber loss 30, 32
human brain 2, 16, 43
hyperparameter tuning 23, 112–13, 119, 123–4
hyperparameters 22, **23**, 32, 33, 36, 76, 86, 88, 118, 122–5, 134–5, 154

I

image captioning 45
image classification 17, 45, 48, **57**, 71–2, 90, 119, 133, 136–7, 141, 157
ImageNet 17, 75, 86
inception 66–7, 69, 71, **72**, 86
inductive 82, 87–8
inference 13, 142, 145
information loss 51, 73, 79

J

Jupyter 14

K

Kappa score 157, **162**
Keras 13–14
kernel **23**, 48, 51–4, 56, 71–2, 74, 82, 132–3, 136
knowledge distillation 134

L

latent space 76, 119
layer tuning 124
Leaky ReLU 26–7, 165
learning rate 22–3, 29, **30**, 34, 39, 87, 114, 123, 125, 127–8, 130–1
life cycle 5–6
likelihood 81, 158–60
local minima 31, 33–4, 72, 123, 125, 128
log scaling 122
logarithmic loss 160, **62**
loss function 13, 23–4, 26, 29–34, 63–4, 74–5, 77–8, 116–17
low-level features 4, 48, 55, 72–4, 84
LSTM 48, 79–81, 133

M

majority voting 107, 109
mapping function 6, 11, 21, 106
marginal loss 75
matrix 4, 13–14, 23–4, 50–4, 74, 82
Matthew's correlation 159
Max rule 109
maxima 125, **126**
mean absolute error 30–1, 159, **162**
mean squared error 10, 21, 30, 77, 160, **162**
meta-classifier 107–8
meta-learning 119, 133
minima 27, 29, 31, 33–4, 38–9, 66, 72, 123, 125, **126**, 128–9
mixture of experts 108–9
MobileNet 70, 139
model error 10, 18–19, 29–30, 32–4, 37, 47, **56**, 61, 66, 77, 88, 103, 105–6, 119, 125, 147–9, 154–6
model training 6, 8, 13, 22, 33, 35–40, 47, 66, 87, 89, 97–9, 102, 106–7, 112, 114–15, 118–19, 130, 140, **142**, 145, 155
momentum 33, 123, 127, 129–31
moving average 126–27, 129–32
multi scale 66, 102, 113, 120
multidimensional data 50
multi-layer perceptron 46, 61–2

N

NAS 132–40
negative transfer 90

neuron 4, 16, 18, 20, 23–8, 32, 42, 47–8, 56, 59–61, 66, 71, 73, 121, 123–4, 130
neuroscience 16, 59
NLP 46, **57**, 82, 86, 89–90, 101–2, 130
normal distribution 7, 27
normalization 34, 49, 60, 71, 113, 120–4, 129, 139
NumPy 14, 50

O

one-shot learning 89, 115, 137, 140
optimization 23–4, 34, 50, 64, 66, 99, 123, 125, 127–33, 135, 137–9, 143–4
optimal 4, 10, 29, 34–5, 37, 51, 59, 63–4, 91, 94, 96, 110, 112, 123, 125, 127–9, 132, 134–5
outlier 6–7, 19, 31–2, 112, 116, 122
overfit 8–11, 21, 32–8, 50, 62, 67, 69, 72, 87, 90, 112, 115–17, 119, 123–5, 132, 137, 153, 155

P

padding 51
Pandas 14
parameter sharing 47–8, 53, **56–7**, 98–9, 115
PC-DARTS 138–9
perceptron 16, 44, 61–2
pipeline 6, 70, 134, 141, 145
pointwise convolution 69–70
policy 93–4
pooling 43, 49–51, 56, 58, 62, 66–7, 69, 72–3, 103, 105, 133, 135
precision 49, 103, 150–1, **162**
prediction 4, 17–19, 23, 40, 46, 61, 74, 79, 90, 103, 105–10, 113–14, 122–3, 134, 137, 142, **149**, 151, 153–4, 160–61
prediction variance 103, 106
pre-process 5–6, 14, 78, 112–13
pretrained model 14, 84–6, 88–90, 114, 139
privacy 96, 98–102, 141
probability averaging 109
problem-solving 11, 91, 93
pruning 36, 170
PyTorch 14, 98, 145

Q

quantization 134

R

recall 82, 150–2, **162**
regression 11, 17–21, 25, 30–2, 35–6, 38, 42, 44, 105, 115–16, 119, 128, 159, **162**
regularization 21, 23, 32, 35–8, 50, 69, 106, 115–17, 119, 124

reinforcement learning 91–6, 134–35, 137
ReLU 22, 24, 26–8, 40, 49, 56, 60, 64, 71, 85, 158, 165
ResNet 64–6, **72**, 86
residual block 41, 64, 66
reward 91, 93–6
RMSprop 130–1
RNN 43, 45–8, 56–8, 62, 79–81, 134
robust 31, 73, 75, 103, 107, 110, 118, 132, 141, 144, 145
ROC 152–4, 158, **162**
ROI 58

S

scheduler 94, 129
search space 125, 128, 131–40
security 99, 140–1
self-supervised 56, 77
sensitivity 23, 139, 148, 152, 154, 158, 159, **162**
sequence transduction 78
sigmoid 20, 24, 27–8, 39–40, 42, 60, 71, 165
skew 7, 113
skip connection 64, 66, 132, 134–6, 139
SoftMax 20, 26, 28, 56, 69, 71, 82, 85, 134–5, 138–9, 165
spatial patterns 45, 48–50, 54, **57**, 58, 62, 69, 72, 74–5, 82
specificity 139, 152, 154, 158–9, **162**
speech recognition 3, 5, 45, 47, 89
stacking 48, 53, 76, 81–2, 103, 106–8, 114, 120, 133
stock forecasting 12, 45, 142
stride 51–2, 55, **72**, 134
supervised 11–12, 21, 61, 77, 91, 93

T

tanh 24, 28–9, **56**, 60, 71, 80, 165
temporal dependencies 48
tensor 13–14, 50–1, 138
Tensor Hub 14
TensorFlow 13–14, 50, 98
test error **9**, 34
test set 17, 21, 33, 116, **162**
threats 2, 140–1, 143–5
tool stack 12
trade-off 8–11, 99, 123, 125, 141, 148
train error **9**, 34
training loss 35, 66, 87
transductive 78, 87–8
transfer bound 90
transfer learning 84–5, 87–90, 93, 114
transformation 4, 61, 79, 113, 120, 135
transformer 79, 81–3

U

underfit 6–10, 32–5, 37–8, 62, 116, 119, 123, 125, 153, 155
unlabeled data 12, 76, 88–9
unsupervised 11–12, 76, 87

V

validation error 33
validation set 33, 117, 134, 154–6
vanishing 23–6, 28–9, 38–41, 47–8, **56–7**, 63–4, 66–7, 69, 71–2, 133, 165
variance 6–11, 23–5, 27, 30, 33–5, 37, 81, 103, 105–7, 113, 116, 120–3, 125, 132, 154–5
vector 13, 18–19, 30, 38, 45, 50, 59, 62–3, 73–5, 82, 121, 128–9

VGG 71–2, 86
vision transformer 82

W

weight averaging 106
weight Initialization 23–4, 38, 40, 140

X

Xception 69, 86

Y

Youden's index 158

Z

zero-shot learning 89

For Product Safety Concerns and Information please contact our EU
representative GPSR@taylorandfrancis.com
Taylor & Francis Verlag GmbH, Kaufingerstraße 24, 80331 München, Germany